PRAISE FOR

HOGs in the Shadows:
Combat Stories from Marine Snipers in Iraq

"A riveting view behind the sniper rifle."

—Marco Martinez, author of *Hard Corps*

"Will give you a personal look inside the mentally and physically demanding world of the marine sniper in combat, almost to the point where you can feel the recoil yourself."

—SFC Frank R. Antenori, U.S. Army Special Forces (Ret.), and coauthor of *Roughneck Nine-One*

HUNTERS

U.S. SNIPERS IN THE WAR ON TERROR

Milo S. Afong

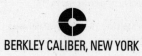

BERKLEY CALIBER, NEW YORK

THE BERKLEY PUBLISHING GROUP
Published by the Penguin Group
Penguin Group (USA) Inc.
375 Hudson Street, New York, New York 10014, USA
Penguin Group (Canada), 90 Eglinton Avenue East, Suite 700, Toronto, Ontario M4P 2Y3, Canada
(a division of Pearson Penguin Canada Inc.)
Penguin Books Ltd., 80 Strand, London WC2R 0RL, England
Penguin Group Ireland, 25 St. Stephen's Green, Dublin 2, Ireland (a division of Penguin Books Ltd.)
Penguin Group (Australia), 250 Camberwell Road, Camberwell, Victoria 3124, Australia
(a division of Pearson Australia Group Pty. Ltd.)
Penguin Books India Pvt. Ltd., 11 Community Centre, Panchsheel Park, New Delhi—110 017, India
Penguin Group (NZ), 67 Apollo Drive, Rosedale, Auckland 0632, New Zealand
(a division of Pearson New Zealand Ltd.)
Penguin Books (South Africa) (Pty.) Ltd., 24 Sturdee Avenue, Rosebank, Johannesburg 2196,
South Africa

Penguin Books Ltd., Registered Offices: 80 Strand, London WC2R 0RL, England

The publisher does not have any control over and does not assume any responsibility for author or
third-party websites or their content.

Copyright © 2010 by Milo S. Afong
Cover design by David S. Rheinhardt
Cover photo by Amatangelo Pascuiti
Book design by Tiffany Estreicher

PRINTING HISTORY
Berkley Caliber hardcover edition / June 2010
Berkley Caliber trade paperback edition / June 2011

Berkley Caliber trade paperback ISBN: 978-0-425-24112-7

The Library of Congress has catalogued the Berkley Caliber hardcover edition as follows:

Afong, Milo S.
    Hunters : U.S. snipers in the War on Terror / Milo S. Afong.
        p.   cm.
    ISBN 978-0-425-23436-5
    1. Iraq War, 2003—Personal narratives, American.   2. Afghan War, 2001—Personal narratives,
American.   3. Snipers—United States—Biography.   4. Snipers—Iraq—Biography.   5. Snipers—
Afghanistan—Biography.   6. United States—Armed Forces–Biography.   7. War on Terrorism,
2001—Personal narratives, American.   8. Psychology, Military—Case studies. I. Title.
    DS79.764.U6A34 2010
    956.7044'34—dc22   2009050687

PRINTED IN THE UNITED STATES OF AMERICA

10   9   8   7   6   5   4   3   2   1

To Bradford Afong.
Thanks for everything, Pops.

CONTENTS

CONTENTS

INTRODUCTION

"Wars may be fought with weapons, but they are won by men."

—**General George S. Patton**

OVER the years, I've realized that most people are fascinated by snipers and the art of sniping. I think it's because people are awestruck with the way that snipers kill; methodically and precisely. I imagine that for the average person, taking another human's life is beyond comprehension, and that's why, quite often, people ask, "Have you ever killed anyone?" or "How many guys did you kill?" or "What is it like to kill someone?" These questions are a doorway into a world unfathomable to most.

Recently however, information on snipers has been revealed much more frequently. With media coverage, multiple books, and instant access to information, the image of snipers has shifted. People understand now, more than in the past, that snipers are not stoic, bloodthirsty killers, but exceptional individuals who work hard at a job most people wouldn't want to do. Much of this information highlights the unprecedented stories of modern snipers in battle, but impressive tales of snipers are not something new.

The unique history of snipers in combat has cultivated legendary tales, such as the duel between Russian sniper Vasily Zaytsev against the German sniper Heinz Thorvald in the city of Stalingrad during World War II. Let's not forget one of the most respected Marine snipers ever to use a rifle, Chuck Mawhinney, in Vietnam, with 103 confirmed kills and 216 unconfirmed. There are other examples of selfless snipers who knowingly sacrificed their lives—reports like that of Army Special Forces snipers Gary Gordon and Randall Shughart, who went to the aid of a downed helicopter surrounded by hundreds of militants in Mogadishu, Somalia, in 1993. Both men were posthumously rewarded with the nation's highest military honor, the Medal of Honor.

Today, snipers in this War on Terror are also gaining recognition. Since the United States was forced into this conflict with a new enemy, we've learned that this enemy is fanatical and beyond negotiations. He has no regard for innocent lives and hopes to use fear as his primary weapon. Surely, no matter how we combat him it is always a struggle, but judging from all accounts, there is a weapon we possess that cripples and terrifies him. This weapon is a man with a long rifle, known as a sniper.

The following pages detail the real-life stories of snipers in the War on Terror. From service to service, these snipers may wear different uniforms, use different weapons, and have different customs, but their mentality and the result of their actions are the same. In this war, the military sniper has moved ahead by leaps and bounds in his tactics and equipment, but it is still the man behind the rifle who must face his enemy head-on. In this new chapter in history, it would be a shame if the lives and actions of such warriors went unwritten. This is the reason for *Hunters*.

ONE

PRE-DEPLOYMENT

IN the War on Terror, no other weapon strikes more fear into the enemy than a sniper. From the crowded streets in Iraq to the dusty cliffs of Afghanistan these masters of precision are extremely lethal against any fighter and in any environment. In Iraq, al-Qaeda members, fearful of U.S. snipers' ability to strike from nowhere, have issued orders to target snipers first, while elsewhere snipers have completely demoralized entire militias by isolating and neutralizing enemy fighters hiding among large crowds. In Afghanistan, long-range precision fire has brought death to the doorstep of Taliban and other fighters who believe that sheer distance is protection.

The operators in today's U.S. Army, Navy, and Marine Corps sniper communities are the best. Earning the title of sniper in each service is no easy task, and those who are snipers have been tried and proven before receiving the privilege of employing the esteemed long rifle. But what does it take to become a member of this prestigious profession, and what skills are needed to operate as a military sniper?

Surprisingly, there is no specific mold for an individual in this trade. Often, people assume snipers to be backwoods folks born with a rifle and raised on hunting. Sure, there are a lot of country boys who become good snipers, but there are also plenty of city boys who wield the long gun, and good ones at that.

Ultimately, snipers are a collective mix. They originate from all walks of life and are as different as different can be, but what allows these individuals to become snipers is certain similar inherent qualities. Military sniper programs look for these traits before teaching individuals the trade and deploying them in combat.

Qualities Needed

Just as in any elite organization, sniper programs look for individuals with exceptional abilities before accepting them into the community. Potential sniper candidates need certain qualities, not only for success in some of the hardest training that the military has to offer—sniper training—but also to flourish in real world combat operations. During screening, selection, and sniper school, candidates are examined for these particular qualities.

By far the first and most fundamental quality needed to become a sniper is heart. Heart is the drive behind never quitting under any circumstance. Unfortunately, this quality cannot be taught, but it is highly sought after in individuals. In the infantry, when units hold selections and senior snipers examine new candidates, they search for this trait right away. During physical training it may seem that those who finish first will be selected. It will happen, but senior snipers would rather choose individuals who try their hardest all the time. They know that physical conditioning can be improved, shooting and tactics can be taught, but a person either has the inherent will to succeed or doesn't. Period.

Another quality needed is self-discipline. This involves many aspects but the first is good behavior. Sniper units screen individual service record books for bad behavior. This can automatically disqualify a candidate. The second aspect of self-discipline is one that every sniper must continually improve, and that is great physical conditioning. One Marine sniper, Sergeant Joseph Morales, explains why this is one of the first qualities needed:

It was my first deployment in 2000, and I was a boot. I wanted to become a sniper as soon as I joined the unit, but I knew that I needed more training and experience. I knew that it was going to be physically demanding when I heard senior Marines, who weren't even trying out, talking about all the strenuous activity involved.

Within two weeks I knew what they were talking about when I found myself covering endless amounts of distances and scaling a fifty-foot cliff with an eighty-pound pack on my back and my weapon in one hand. This definitely pushed my body to the limit, and it was just the beginning. I figured that if I was to become a sniper I needed to train my body to endure anything.

Intelligence is also desired and is needed to learn the tremendous amount of information presented to snipers. Good judgment and common sense also come into this equation. Operating several different communication devices, cameras, and high-powered optics; knowledge of ballistics, ammunition, range formulas, rifles, the procedures for calling supporting arms and close air support, land navigation, mission planning, collecting intelligence, and enemy weapons capabilities are just some of the job requirements.

Two types of maturity are also important for a sniper. Emotional maturity is needed to cope with the strain and violence associated with sniping. It takes strong individuals to constantly risk their lives and set aside their emotions while considering factors such as death and physical pain. Professional maturity is just as imperative. Snipers must be able to report to and brief commanders in several areas. Tactical employment, full use of a team's capabilities, mission planning, coordination with supported units, and directing air or ground support are essentials. In doing any of these activities, a sniper represents his team and its capabilities in the highest fashion. This gains trust for further operations. On the flip side, if a commander loses trust in his snipers, it is not unheard of for him to cease all sniper operations.

Mental stability is also needed to deal with killing. Units screen against empty-headed and heartless killers and also for those just the opposite, troops who are deemed too soft for the job. Programs do not accept those who believe that if they were behind a sniper rifle, they'd kill everyone in sight. That is not the job of a sniper and that mentality is dangerous to everyone.

On the other hand, killing is part of the job. Fundamentally, the role of a sniper is to deliver death—meaning that the man behind the rifle has to squeeze the trigger when it is called for. The power to take another man's life is a heavy responsibility, and snipers must be mentally prepared for the act.

Finally, potential snipers need to be proficient with the service rifle. Clearly, sniper units only accept individuals with expert rifle qualifications. This demands that potential candidates have a strong grasp of the basic shooting techniques needed, so that more advanced methods can be built on.

These qualities are just a few of many that potential snipers should have if they want to learn the trade of sniping.

Learning the Trade

Individuals wanting to become snipers are drawn to many aspects of the trade. The independence of small teams and the ability to be self-reliant are huge reasons people go into this field. Being able to target the enemy with little chance of detection, use advanced weapons and equipment, and learn superior tactics are also benefits. Whatever the motivation may be, potential snipers should be aware that learning the trade is a long, tough journey.

In the world of Special Operations, sniper qualification is merely one skill among many. The U.S. Army, Air Force, Navy, and Marine Corps each have particular units that fall under America's tip of the spear, the United States Special Operations Command or USSOCOM. Within this group, each unit employs its own snipers and what is distinctive about being a qualified sniper here is that the men are able to use their skill when the situation is called upon, but they aren't strictly limited to sniping alone.

U.S. Army Special Forces soldiers, or Green Berets, are excellent soldiers *and* snipers, to say the least. Passing the U.S. Special Forces qualification course is next to impossible, and operating as a member of an ODA or Operational Detachment Alpha, Special Forces A-team member is even harder, so to become a sniper among one of these teams is truly exceptional. To become one, an SF soldier must have consistent expert qualifications with a service rifle. From there, a psychological exam is given to ensure potential candidates are mentally ready for the repercussions of the job. As Green Berets are drawn from many fields, some soldiers arrive to an A-Team having already been sniper qualified through the regular Army's basic scout/sniper course.

Those in A-Teams who need sniper qualification attend SOTIC. This is the U.S. Army Special Forces eight-week course known as the Special Operations Target Interdiction Course. In this highly prestigious course, soldiers are taught the basics of sniping without the physical repercussions and consequences enforced in the conventional Army sniper schools. Typically, the students are Green Berets, though the doors have been opened to Army Rangers and regular Joes.

U.S. Navy SEALs also have in-house sniper training. To be considered for sniper training, SEALs must have an expert rifle qualification from the regular Navy's shooting course, usually two deployments as a "team guy" in a SEAL platoon, and a recommendation by the platoon chief or commander. Once the blessing is received, SEALs attend a twelve-week course.

The sniper school is based out of Coronado, California. Here, SEALs learn much of the same skills as those in other services schools. One factor in which SEAL sniper schools differ from their counterparts, though, is the amount of technology involved in learning sniping. With a heavy emphasis in reconnaissance and surveillance, SEAL snipers must master techniques for collecting and reporting information. This is taught in the first two weeks of sniper school, where students learn an in-depth use of cameras and computers along with intricate techniques of employment and enhancement with their tools.

The next ten weeks is broken into two phases. First is the four-week scout phase, where students learn everything from urban hides, to the use of ghillie suits and stalking, as well as traditional sniper skills such as Keep in Memory or KIM games. The final six weeks are when snipers master the art of shooting. This incorporates three different weapons systems.

Another unique aspect that SEAL sniper training covers is mental management. In this instruction, students learn to perform successfully under heavy amounts of stress and in critical scenarios. This proven method was introduced by Olympic gold medalist and eighteen-time World Champion shooter Lanny Bassham.

Becoming a SEAL sniper is one of the hardest routes in the Navy, with a 70 percent fail rate at Basic Underwater Demolition/SEAL training, and another 50 percent fail rate at SEAL sniper school. Very few individuals ever become SEAL snipers.

The operators from Marine Corps Special Operations Battalions, or MSOB, are given the chance to become snipers after finishing the basic reconnaissance course and a few other military occupational specialty essential schools. Special Operations Marines are selected to attend the basic scout/sniper course after having qualified expert with the service rifle, completed at least one deployment, and receiving a recommendation from his platoon sergeant. However, as it is with Army Special Forces, many Marines join the ranks of MSOB having already served as snipers in the infantry.

Marines already sniper qualified in MSOB can attend another sniper course. It is known as MASC or Marine Special Operations Forces Advanced Sniper Course. One instructor explains the difference between this course and the basic scout/sniper school:

> *The basic course teaches basic skills that build a sniper from scratch. Our course takes those basic skills, introduces new concepts, ballistic software, and equipment, and enhances those skills to fit more of a rapid target engagement environment. The techniques we teach are applicable to any environment, long or short range, urban or rural, and although there is training in an urban setting, a good portion of the instruction is taught on the range, shooting at various targets and distances. While the end result of a qualified urban sniper is similar, we do not delve into the surveillance and reconnaissance aspect of sniping. We focus on enhancing the sniper's understanding of engagement techniques, and his proficiency in those techniques.*

It must be said that the role of scout/snipers among the U.S. Army and Marine Corps infantry is slightly different from those in Special Operations. Special Operations snipers have an extended variety of specialized skills to include sniping, but those in the infantry battalions concentrate exclusively on sniping and working in sniper teams.

The first step to becoming a sniper in the U.S. Army and Marine Corps infantry begins at sniper selection. Every few years, new snipers are needed and candidates are chosen after having taken part in a selection. Good eyes, good behavior, a strong body, and an even stronger will are needed to make it through this trial. Sniper selections test the mental and physical aspects of a candidate. The object is to weed out the soldiers and marines who turn up for the wrong reasons. Those who question themselves or doubt their ability should not attend.

For Army sniper Adam P., his battalion sniper selection was four days long.

We were first given a physical fitness test and several candidates were eliminated. We were then given a written test made up of basic common tasks like calling for fire, first aid, and medevac procedures, followed by classes on range estimation, target detection, and other basic sniper tasks. The next three days were a combination of mental and physical challenges, culminating with a nine-mile ruck march with an eighty-pound pack. We started with fifteen men, and by the time we got to the ruck march, we were left with only four. During the march we lost one, leaving three guys, including me.

In the Marine Corps, selection is known as indoctrination and it stresses the same ideas. One sniper opened up about how he oversaw a selection process:

I made it hell on the candidates during my first selection process as a monitor. I wanted as many Marines to quit as possible, to keep the community elite. In the Marines, three words strike fear and misery into sniper candidates. Find A Pole!

It means to immediately, just as the words are said, elevate your feet on anything nearby. It could be a desk, a chair, a windowsill, the hood of a car, a tree, or even a pole. Once there, we commenced push-ups. It doesn't seem like much, but after a week of this happening around thirty times a day, along with pack runs and pool time mixed with brain-jarring tests while under sleep and food deprivation, quitting seems like a dream come true.

When the selection process is through, and if you've made it into the scout/sniper platoon (Marine Corps) or the reconnaissance and surveillance target acquisition platoon (Army), the next step is the infamous scout/sniper school.

Essentially snipers pride themselves on the ability to do

three things well: shoot, move, and communicate. Shooting is the most recognized skill, but without being able to move undetected and the ability to report your activity to higher, shooting is a non-factor. Together, these skills and more are the building blocks taught in scout/sniper basic course.

The U.S. Army Scout/Sniper Basic School is held in two locations. One is at Fort Benning, Georgia, and the other is on Camp Robinson, Arkansas. While the courses teach the same doctrine, there are a few differences. Camp Robinson's course is instructed by Army National Guard soldiers and is broken up into two segments, while the cadre at Fort Benning is made up of enlisted and former enlisted snipers and is five weeks straight.

Both schools, however, cover the same curriculum. Sniper students learn a combination of several field crafts and skills, including land navigation; patrolling; stalking; target detection; field sketch; range estimation; ghillie construction; urban operations; hide selection, construction, and occupation; as well as tracking and counter tracking and much more.

Marines in the scout/sniper platoons earn the title of sniper at one of four basic scout/sniper schools: First Marine Division on Camp Pendleton, California; Second Marine Division on Camp Lejeune, North Carolina; Stone Bay at Quantico, Virginia; and Third Marine Division on Kaneohe Bay, Hawaii. Each of these schools stresses the same curriculum and produces some of the best snipers in the world.

Since 2001, the Marine Corps sniping program has seen many changes, the primary being that the basic course is now eight weeks long as opposed to ten. From this course, students gain a strong understanding of shooting and ballistics, stalking, field skills, and much more. Another change is that the final two weeks are now incorporated with an additional three weeks and are part of the five-week team leaders course. The TL course teaches mission planning, sniper employment, more in-depth training with communications and supporting arms, and other responsibilities directly associated with the TL position. Upon graduation, Marine snipers earn the military occupational specialty of 0317, and just as important, they earn the right to be called a Hunter of Gunmen, or HOG.

Beyond the Books

Clearly, all military snipers learn the skill to become a sniper through their basic courses. Learning, however, does not stop there. Though sniper schools teach the skill, they also teach a handful of essential qualities not learned through books, with the first being teamwork.

Gone are the days of individual snipers operating alone. Nowadays the concept of at least a two-man team has proven to be more efficient for snipers to kill and to survive. In sniper school, students are never solo and one instructor at First Marine Division Scout/Sniper School explains why:

> We implemented the one-arm's distance [students' distance from their partners at all times] rule in order that they may learn accountability, and to ensure they're cognizant of what is going on at all times.
>
> We try to stress good habits in our students. Mentally, it keeps the guys driving on because they can push each other (specifically regarding their type-A personalities). In essence, they try to be better than their partner out of pure competitiveness. Also, learning to work with others and being that "team player," guys gain maturity and experience. This serves as an opportunity to learn other tricks of the trade, or one might have a solution to a problem the other student never considered as well. Lastly, "iron sharpens iron." With that in mind, students not only learn to work with each other, but even teach or disciple others.
>
> Now, we as instructors try to give each student responsibility so that this may be a growing experience for him. Without responsibility, how and when will he learn it? Big boy rules apply here, only because these students will become team leaders. In addition to that, the instructor staff gives each student their all, fully knowing that someday we might operate alongside these Marines. This doesn't mean

the students take advantage of this opportunity, and quite often, they dismiss it.

For the students who are lacking in leadership, we put them in charge. All too often, their peers will help them learn to mature. Otherwise, we focus on those areas where they are lacking, and build on their strengths because teamwork builds character, maturity, endurance, strength, and leadership.

To survive school, and combat, snipers learn to trust their teammates. Of course, in school all individuals must master all of the skills independently, but to survive it all, teamwork is also needed. This is true especially during the shooting phase.

Students must master both shooting and spotting to qualify for known and unknown distance shooting. Each must trust in his partner's ability to provide exact calls for wind, distance, and second-round adjustments to hit the target. If this ability is not there, the spotter will fail the shooter, and the shooter will fail sniper school.

Known distance is the easy aspect of sniper shooting. To pass the known distance portion of First Marine Division Scout/Sniper School, students must hit at least twenty-eight of thirty-five targets starting from three hundred and ending at one thousand yards. Effective communication is necessary, as students will be tested with moving and bobbing targets at six hundred to eight hundred yards, and exact wind adjustments are needed for students to hit targets at one thousand yards.

Unknown distance is more difficult. During unknown distance training, the shooter gives the spotter his target's dimensions and the spotter is responsible for calculating the distance through formulas. To pass, the team must trust that each person knows his job and can properly adjust the shooter's round should he miss.

Unknown distance qualification is based on a point scale at the U.S. Army Scout/Sniper School in Fort Benning, Georgia. To pass, students must receive 70 percent or better. With ten targets, students are given ten points for a first-round hit and five points for a second-round hit. This starts with two minutes to range a target using a scope. After the two minutes, students have ten seconds to engage. If they miss, they receive an extra ten seconds to re-engage. To complete this program, the team needs rapidly to apply formulas and wind and elevation adjustments, and to trust each other's abilities.

In sniper school, students are also taught the rewards of perseverance. One sniper mantra defines this perfectly: *suffer patiently, and patiently suffer.*

Suffering is the name of the game for snipers, and during school, students learn this quickly. Bad weather, long movements, hunger and fatigue, equipment malfunctions, hassle from the instructors, and generally any type of setback is to be expected, but the best thing a student can do is take it all in stride. The concept is that suffering in training leads to success in combat, and that adapting to any situation leads to overcoming the obstacle.

Instructors teach this by purposely pushing students to their limits, and the reason is twofold. First, the instructors only want to qualify those who really want to be there. The stress during school is needed to make those who do not truly want to operate quit, because the worst possibility for a sniper team in combat is to have a teammate who does not want to be there.

Second, men who've been pushed to their limits and succeed gain heaps of confidence, and to be a sniper you need just that. Snipers must be certain that they can accomplish any mission, not only for their survival, but also for the success of the supported unit's mission.

School also teaches individuals to think independently. The problem in this area is that sometimes the proper use of a sniper team is not determined by the sniper teams themselves, which can lead to poor employment. For snipers, the mentality is to stay one step ahead of their enemy in every way. For this to happen, sniper team leaders need the freedom to make decisions that are vital for success. If that freedom is not established among the supported unit or the battalion, then the team is already at a handicap in its capability to be instrumental.

Another quality students learn from sniper school is patience. This is especially needed during the stalking phase.

The experience gained from here then spills over into other aspects of sniping.

Stephen Johnson, a Marine sniper, describes his experience:

The aspect that stands out the most from my learned patience from stalking, in sniper school, is the marksmanship aspect at the end of the stalk. Even though we shot blanks, I still remember the difficulty in assuring that everything was correct. I remember the mental checklist—did I camouflage well, is my blast lane long enough, can I burn through the vegetation to confirm sight of the objective, shadows, deflection—all running through my head simultaneously.

This aspect of training I found to be absolutely true when confronted with an insurgent placing an IED alongside MSR bronze near Haditha, Iraq. I remember that day vividly.

It was clear and crisp with a steady wind reaching gusts of twenty miles per hour. It was the middle of the afternoon when a white truck stopped at the road intersection and two males got out. This immediately drew my attention as I sighted in behind the scope to get a better view of what they were doing. They couldn't have stopped at a better spot in relation to where I was set up. It was almost as if they ended up directly in my line of sight.

Unfortunately for them the driver of the truck jumped into the back of the truck bed and picked up something heavy. I could tell by the way his back was slumped over,

*and he struggled to hand it to the passenger waiting along-
side the truck. He received it and ran to the side of the road.
This immediately led me to believe that they were planting
a roadside bomb.*

*From my training and experience I knew I had to act.
I tracked the suspect through my scope until he stopped
along the edge of the road. This would be my one opportunity. All my training began to race through my head as I
refused to accept missing a shot. I lined up the crosshairs on
his chest silhouette and applied a slow and steady trigger
squeeze. He immediately hit the ground as I saw a pink
mist spray from his body.*

*Just as during stalking, my final firing position was
undetectable. The patience needed for making this shot
transferred from training to real life and was successful.*

Marine sniper Jon S. also describes two very similar situations, one during school, the other in Iraq.

*During sniper school, my final objective was to make two
shots at an unknown distance. My partner and I waited in
position over twelve hours to make the shots. We needed
every ounce of energy to stay awake because we'd been
moving for four days straight, constantly going from one
mission to another with resupplies in between. By the final
day, we were exhausted. As we waited, the order to shoot*

*was finally given over the net, and if we'd been compla-
cent, we would have failed.*

*In Iraq, during Operation Phantom Fury in Fallujah
2004, our team had been awake for over twenty hours. Insurgents continued to flank the company we were supporting and
our team was asked to over-watch for a platoon moving to
secure an area. Everyone in our team was exhausted, but
when we were in position, my spotter noticed three men with
weapons moving towards the Marines in front of us and we
eliminated them. After we killed them, I realized how the
situation in sniper school mirrored actual combat.*

Snipers also need patience because not every mission under-
taken will produce a kill. The chance to kill can take anywhere
from weeks to months, or it may never be presented. In the same
way, patience is required for a sniper to make the best decision
on how to kill the enemy. Very rarely does the enemy travel
alone, either in Iraq or Afghanistan. For snipers this means using
the perfect weapon to strike in order to increase the maximum
body count. Often, snipers make the difficult decision to deny
themselves the use of the sniper rifle and to rely on supporting
arms. This may sound logical to most, but for a trained sniper it
is a heavy sacrifice not to engage with the sniper rifle.

These skills and qualities produce men who become force
multipliers for their units. They also give the snipers the tools
they need to carry out their primary function—to kill the enemy.

Killing

For military snipers, killing is their purpose, and it is crucial to their survival. Militaries have recognized the need for a sniper's precision, and there is no other time more crucial for accuracy than the Global War on Terror. Snipers have proven to be the perfect solution in environments where civilians are present and collateral damage needs to be minimized.

The act of actually sniping someone in combat is thought to have something mystical about it. Any combat sniper, however, will tell you that there is no aura during the kill; it is just a squeeze of the trigger, and nothing else. It is common for snipers to experience the nervous feeling—buck fever—during their first kill, stemming from the fact that they are actually about to fulfill the goal of all of their training.

By and large, the will to kill is ingrained in most snipers before they reach sniper training. Since sniper communities are made from the infantry or Special Operations units, the men have learned to kill from the start of their training. By the time sniper training is completed, taking the life of an enemy is simply the job they've been assigned to do.

One U.S. Army sniper instructor says:

Sniping is first and foremost about killing people. If one does not think that this is something they can handle, then sniping is not for them. Many people say that killing a man is the worst thing they have ever had to do. This may be true for them, but for me and most snipers I know, it is only a job. Killing the enemy is a task given, and snipers execute this task because it must be done. For most that I know, it ends there. There are images and things that will stick with me for the rest of my life, but pulling the trigger on an armed enemy trying to harm friendly troops is not something that will bother me now, or ever.

I, and all the snipers I know, agree with this mentality. I've been asked, "How do you deal with killing people?" My reply is that I know that everyone I've killed was an enemy combatant, and was intending to harm friendly troops or myself, and I can live with that. I believe that some snipers may have a hard time dealing with killing if there are uncertainties involved.

Tools of the Trade

The equipment and weapons used by snipers in the military differ between services and units. The common factor among all, though, is the primary use of a 7.62mm NATO round. Each unit also uses heavier sniper rifles for farther distances and armor penetration. Night vision, thermal imaging, radios, global positioning devices, high-powered optics, periscopes, tripods, night sights for the sniper rifle, laptops, cameras, and suppressors are the typical equipment available for sniper teams.

In the conventional Army, scout/snipers have three weapons available for use. The first is the M24 Sniper Weapon System or SWS. Many snipers consider this their bread-and-butter weapon for its proven reliability in combat. This is a bolt-action, 7.62mm rifle made by Remington and is very similar to the M40 series that Marine snipers use, with some minor differences such as stock and scope. The longest recorded shot in Iraq was 1,250 meters (4,000 feet) by an Army sniper with this weapon.

Next, Army snipers have the M110 Semi-Automatic Sniper System or SASS. Knights Armament, a company with an extensive history of developing exceptional weapons, designed this weapon as the next best thing for Army snipers. Since 2007, this weapon has been fielded by Army snipers in combat with the intention of replacing the M24.

This semiautomatic rifle fires a 7.62mm NATO round and comes with a suppressor and an attachable night sight. It is effective out to 800 meters (2,600 feet), with the capability to shoot out to 1000 meters (3,300 feet), sniper dependent. Though it is an exceptional weapon, some snipers feel that it should not be replacing the very reliable M24. Instead, it should be used in conjunction with it as both rifles have different applications in which they are useful.

The last rifle available is the M107 Long-Range Sniper Rifle, LRSR. Made by Barrett, a company specializing in large-caliber rifles, this semiautomatic, .50-caliber sniper rifle seems to conjure mixed feelings. Some feel that it is too heavy and cumbersome and not as inherently accurate as they would like. It does, however, have light armor–penetrating capability, which makes it extremely useful in defense. This weapon also has the ability to cover particularly long ranges, such as in the mountains of Afghanistan.

Just like any other sniper, U.S. Army Special Forces soldiers select weapons depending on the mission. The sniper rifles they can choose from are the same as conventional Army snipers with the exception of a more powerful rifle. The bolt-action .300 Winchester Magnum that they use is a Remington model 700 inside an Accuracy International stock with a variable powered scope. The rifle is known as the MK-13.

Marine snipers have a few weapons of their own. The primary choice is between the M40A3 and M40A4, which have the same characteristics except that the M40A4 has a detachable, five-round magazine. These rifles are a modified Remington model 700 with a Schmidt and Bender scope and use 7.62mm NATO rounds.

Another weapon is the MK-11. This rifle is a 7.62mm, semiautomatic, ten round–capacity rifle that looks similar to the M16. It also has a suppressor for noise reduction, but some snipers claim this makes the chance of malfunction greater.

Finally, Marine snipers also have the M107 .50-caliber sniper rifle. Some have the same complaints about the weight, but many Marine snipers love the penetrating capabilities of this heavy rifle.

Each U.S. Navy SEAL sniper has a suite of rifles. They have the MK-11, also used by Marine snipers; the MK-13, also used by Army Special Forces snipers; a modified M14, using a 7.62mm NATO; modified M4s; and a heavy, single-shot, bolt-action, .50-caliber sniper rifle.

The Complete Package

These days, the typical sniper deploying in the War on Terror is well trained, and very deadly. This is a result of exceptional training incorporated with the experiences of each individual. Snipers know that basic scout/sniper school and mastery of weapons are the foundation, but further development and training are very necessary before deploying.

To add to their arsenal of deadly knowledge, snipers can also attend additional training. Urban sniper courses, high-angle shooting packages, survival/evasion/resist/escape courses, and other advanced sniper programs are available. Individual units also have training in place that prepares teams for simulated combat environments before they set foot in country. These courses give the deploying snipers a mental confidence over most.

This confidence is essential to a sniper's survival. U.S.

snipers deploying to combat are determined, and one U.S. Army sniper, with extensive combat experience, reveals the mentality that one needs:

I believe that a sniper should go to war 100 percent ready to execute his mission, no matter what that may be. He must have the mind-set that he is in complete control of himself and any men he may be in charge of and must be prepared to put himself in harm's way in order to accomplish his mission. He also must have total confidence in his equipment and his ability to use the equipment. Snipers know that the mission they are required to do is more dangerous than most, and most snipers thrive on the challenge.

I, personally, had a strong desire to eliminate the enemy. Setting up a perfect ambush and catching the enemy by surprise is a feeling that nothing else can rival. When I deployed, I decided that I would do all that I could do to kill as many of the enemy as possible. It's not because I was bloodthirsty or anything. It was because the more of them I kill, the fewer there are to shoot at or try to blow up my buddies in the line units.

TWO

AREA OF OPERATIONS: AFGHANISTAN

In Afghanistan, the enemy, the hard corps Taliban and al-Qaeda fighters, they're not scared
of us. They're just as excited about killing an
American as we are about killing them, but I
can tell you that they fear the unknown.

—U.S. Special Operations sniper, 2009

IN the winter of 2007, twenty-three-year-old U.S. Army
sniper Sergeant Tyler Juden waited patiently for his prey. His
team remained cloaked high in the Spin Ghar Mountains of
northeastern Afghanistan, and from their hilltop, Tyler kept
an eye out for movement below. He was not searching for
quail or mule deer like back home on the Kansas plains. Here,
in these mountains, the prey was Taliban insurgents.

The sniper team observed into the Khyber Pass, a strategically important stretch of ground bordering Afghanistan and

Pakistan. The winding valley allowed Coalition Forces to ship vital supplies from Pakistan, but it also allowed enemy fighters to attack Tyler's base and slip back across the border, which they'd been doing.

The attacks happened every Wednesday morning at 0900. Tyler's small outpost took mortar fire for a month before his team was dispatched. Unfortunately for the enemy, their pattern of attacks, along with a lightweight counter-mortar radar, gave the U.S. soldiers a fix on their firing position.

Planning took days. Later, Tyler, his spotter, a forward observer, and a SAW gunner moved from their base. It was after midnight. They hitched a ride with an infantry patrol until they were about seven kilometers (four miles) from the target area. The route to their destination had jagged cliffs and rugged terrain, allowing only foot patrols to enter the area. That seemingly gave the Taliban an advantage, but in the days to come they would no longer be safe there.

For more than two days, the four-man team lay on the mountainside. Tyler and his spotter were fifty yards above the forward observer and SAW gunner. Their weathered Army Combat Uniforms (ACUs) matched the sparse brush and rocks, and the painted M110 blended perfectly with the dirt. A skillful map study had put them on the highest point in the area. From that position they had a perfect view of the canyon and ravines below them. The radar indicated that the mortar team would fire within a few hundred yards of them. When they did, the snipers would be waiting.

On Wednesday morning at 0900, the soldiers scoured the valley for the enemy mortar team, but found nothing. At 0915, Tyler began to question the intelligence they had received. When 0930 rolled around, all was still quiet, until Tyler's partner spotted something.

Three men strolled into view on the other side of the valley. Tyler's spotter quickly found the range to targets; it was six hundred meters (two thousand feet) and counting. The snipers knew right away that these were bad guys. Besides the fact that they had AKs and green chest rigs, their Puma tennis shoes were a dead giveaway. The shoes matched those used by other enemy fighters that the soldiers had targeted before. Tyler notified the forward observer, who also saw the men. He immediately contacted two Apaches already in the area and had them start in on an artillery support mission.

Atop the hill, the snipers prepared to engage. Typical in Afghanistan, Tyler's shots would be difficult ones; a breeze swept over the mountain behind them and met a crosswind from left to right at five hundred meters away. In addition to that, the angle had Tyler aiming downhill, plus the closest fighter, the tail man, was more than six hundred meters away. Before engaging, though, Tyler and the forward observer decided to coordinate the attack to mask Tyler's shot behind the impact of artillery. The explosions would keep their position concealed in case more fighters were near.

Minutes passed and the artillery did not show. Tyler was forced to engage before the targets moved out of range. By now the men were at six hundred and eighty meters, and Tyler corrected his scope's elevation and windage. He had been on plenty of deer hunts and won marksmanship contests in high school, but this was the real deal. Four years of sniper training, sniper school, all the "what ifs" and "could I's" came down to the following few seconds. After a deep breath, Tyler's natural point of aim put his crosshairs' center mass on the tail man. Everything felt right and he let the bullet fly.

Traveling at 2,571 feet per second, the round tore through the man's rib cage and instantly dropped him. In front, his friend turned and stared in confusion while Tyler slowly squeezed the trigger to shoot him also, but in mid-squeeze an artillery round blasted the mountainside two hundred meters away from the three.

The explosion scared the enemy off. As they ran, Tyler concentrated with all his might on hitting one more, but his first couple of shots were off. He had only one chance left. More than seven hundred meters away, his target ran at a dead sprint, quickly closing Tyler's window of opportunity. Then his training kicked in, and he sighted ahead of the man to use the ambush method. Just when the man hit his mark, Tyler shot and recovered in time to see the outcome. His bullet hit just behind the man's rear leg, so close that if the man had kicked his leg up a little higher, he would have been hit, but sadly for Tyler, he slipped away.

By now, the Apaches were en route, directed by the forward observer. Tyler traversed back to the man he had already shot. The man was on his hands and knees, still alive.

"Shoot him again," said his spotter, but Tyler didn't want to. Minutes later, after the man started crawling, Tyler changed his mind.

His windage was slightly off for the second shot. It hit the man's butt and knocked him onto his stomach. Unbelievably, the man sat up moments later to rest against a rock. Tyler finished him off with one last shot, which penetrated his neck, above the collarbone. Just then, the Apaches arrived and eliminated the two men who had been running. Later, a patrol arrived and inspected the bodies and collected intelligence from them. In the end, the soldiers' forward operating base didn't receive any more indirect fire.

In the past eight years, missions like these have become standard for snipers in Operation Enduring Freedom. In the quest to defeat al-Qaeda in Afghanistan and to overthrow the Taliban, U.S. snipers have played a variety of roles, encompassing offensive, defensive, and conventional sniper operations. For the first few years of the Global War on Terror, however, some snipers dreaded deploying to Afghanistan, considering it to be dull. The first of them had arrived after the bulk of the fighting to find that the enemy had been killed, captured, or driven into the most remote areas in the region. That would change in the years to come.

Recently, Afghanistan has become a sniper's Mecca. The environment is ideal for snipers hoping to test their skills and training. This no-man's-land introduces many variables that snipers must consider: dangerous landscapes, sprawling distances, unforgiving weather conditions, and a resourceful and hardened enemy fighter.

As it pertains to snipers, defining the enemy can be confusing. Allegiances and pacts between tribes and groups can switch in a second, turning friends to foes and vice versa. There are a few certain enemy fighters though, one being the Taliban.

The Taliban is recognized as "the students." These men implement a strict interpretation of Islamic law and brutally enforce it. Afghan Pashtuns, also traditionally Sunni Islamists, are the bulk of the group, but Pakistani Pashtuns and others have also joined the ranks. Pashtuns have dominated Afghanistan for hundreds of years. Tribal alliances cross geological boundaries into Pakistan, giving the Pashtuns a place to retreat to and rest. The Pashtuns in Afghanistan are conditioned for war and made up most of the Afghan mujahedeen, or *holy warriors*, who fought to prevent Russian occupation in the 1980s. Shorty after that war ended, they formed the Taliban, subsequently taking control of the country.

As of 2009, the Taliban has been relegated to an insurgency, but their influence over civilians and Afghan military is powerful. Geographically, their control covers most of the south and east, but attacks have spread throughout other regions of the country. Like the insurgency in Iraq, Taliban fighters hit and run and employ IEDs as well as indirect fire. Though many conduct cross-border operations or fight from caves and mountainsides, there are also groups living and operating from their hometowns and villages, easily blending in as civilians.

Another enemy is al-Qaeda, translated as *The Base*. Created in part by Osama Bin Laden, this international terrorist organization was initially driven into hiding in 2001, but as of late, they have reemerged. As the self-described advocates and headquarters for future *jihad,* or *the struggle,* al-Qaeda fights to drive non-Islamic agencies and their influences, particularly the United States, from Muslim nations with the purpose of installing fundamentalist Islamic rule.

As fighters, al-Qaeda members are a collective mix. These men come from as far as Saudi Arabia, Algeria, Egypt, Uzbekistan, Pakistan, Chechnya, and Jordan but also include local Afghan Arabs. Their grasp of tactics and weapons instructions is taught at internal training camps. In these camps, individuals learn the use of small arms, explosives, intensive guerrilla warfare, and small unit tactics. A fierce devotion to their ideology makes them more fearless, disciplined, and willing to fight to the death than their Taliban counterparts.

Other fighters fall under the term ACMs or Anti-Coalition Militias. ACMs are groups of fighters and tribes who battle Afghan and Coalition troops. They are found in most regions of the country and have proven to be formidable enemies. Though these groups are not particularly involved with al-Qaeda or the Taliban, some do cooperate with them.

In facing the enemy, U.S. snipers have an integral role in the success of combat operations. Large-scale offensive maneuvers have demonstrated the benefit of snipers supporting the infantry, while individual covert sniper missions have been instrumental in demoralizing and destroying enemy personnel. In Afghanistan, snipers are the unknown factor, killing some while striking fear in the rest. However, that does not mean that they are in any less danger. There, a sniper must be strong, patient, and courageous because there is no telling in what situation he might find himself.

THREE

FIRST OF THE FIRST

IAN Baker was in Australia when his life changed. His sniper platoon from the First Battalion, First Marines, part of the Fifteenth Marine Expeditionary Unit (Special Operations Capable) had spent time with the Aussie military patrolling and shooting on a sniper range outside of Darwin. When they finished, the marines were able to do what they loved best, enjoy liberty.

It was Baker's second deployment. At twenty years old, he had spent all three of his years in the Marines within the battalion's scout sniper platoon. Two years earlier, he had graduated from First Marine Division Scout/Sniper School, learning what it meant to be a certified sniper or HOG. By his second deployment, he was one of the most experienced snipers and a team leader.

In Darwin, after training, the snipers were set free on the town. It meant that Baker could gamble, and at a nearby casino he tested his hand at blackjack, winning a few thousand dollars. The next day, his platoon met at a local tavern to help Baker blow some of the money that he had won. It was September 11, 2001.

Inside the tavern, Baker ordered a pitcher of beer before heading for the bathroom. As he did, he noticed the World Trade Center attacks on television.

"What movie is this?" he asked a local, unaware that it was actual footage.

"It's not a movie. You guys shouldn't be here," he said.

Baker stopped and watched a little longer, realizing the situation. Soon everyone in the bar was glued to the television, completely in shock. Shortly afterward, shore patrol from the Fifteenth Marine Expeditionary Unit (MEU) arrived and announced that everyone needed to report back to ship. On the boat, Baker's company commander told them about the terrorist attacks, and the next morning, the ships pulled out of port.

MEUs were just the type of strike force fit for fighting terrorism. An MEU is made up of 2,200 marines and sailors, with artillery, tanks, amphibious vehicles, attack helicopters, fighter jets, and many more elements all built around the main asset, a single Marine infantry battalion. The First Battalion, First Marines, from Camp Pendleton, California, were the current battalion landing team with the Fifteenth MEU, and they had spent twenty-six weeks training in several environments and scenarios to receive the label Special Operations Capable (SOC). Once they did, they were able to fight in every clime and place, and that is just what they expected to do when the news broke about a War on Terror.

After leaving Australia, however, the Fifteenth MEU steamed toward East Timor. They continued with their original schedule to provide humanitarian assistance to the struggling nation. When they arrived, Baker's sniper team climbed onto the rooftop of a clinic to provide security while other marines gave medical assistance to civilians. All the while, Special Operations forces infiltrated Afghanistan to begin Operation Enduring Freedom. With other American forces fighting, the marines itched to get into the fight, and after a week in East Timor, the Fifteenth MEU aimed for the Arabian Sea off the coast of Pakistan.

On ship, Baker learned that his battalion had received a mission. The Pakistan government agreed to U.S. forces utilizing airfields in Pakistan, providing that none be used for staging attacks into Afghanistan. Vital for U.S. air strikes in Afghanistan, Air Force Combat Search and Rescue needed a secure airfield to operate from, and Jacobabad, near the Afghanistan border, was the perfect solution. For the marines of the Fifteenth MEU, it meant that that they and their Maritime Special Purpose Force (MSPF) team would move in to establish a perimeter and provide security at the airfield in Jacobabad.

The MSPF team was elite among the MEU. The special unit combined four elements: command, assault, support, and security. The main effort among them was the assault element, a Force Recon platoon. At the time Force Recon was as close to Special Operations as the Marine Corps allowed. Together the MSPF team normally focused heavily on maritime operations, specifically direct action and reconnaissance. For them, assisting to secure the airfield would be no problem.

Before deploying, Baker's entire sniper team had passed the Marine Corps Special Operations Training Group's urban sniper course. They had practiced shooting from helicopters and from rooftops and setting up in urban environments. They also had learned the intricacies of timed shooting and how to support the Force Recon marines before they raided targets. When the snipers graduated the course, they were designated as the primary MSPF sniper team.

Working with the Force Recon team was enjoyable for Baker. They were professional and extremely good and had the best training. As their supporting sniper team, Baker's men accompanied them on training operations, and Baker himself constantly provided precision fire. He had grown comfortable sending bullets past the men during training, and he knew that if a real situation called for it, he would be able to do the same.

By the time the marines reached the coast of Pakistan, they were anxious to go. They were ready to secure the airfield and waited for the mission to begin. On the night of insert, Baker gathered his team on the ship's flight deck, and they boarded CH-53 Super Stallions, each man carrying one hundred pounds of gear. Although Pakistan was a friendly nation, the threat of terrorism remained high, and the marines carried ammo and explosives. Baker packed a .50-cal sniper rifle, while his assistant team leader had an M40A1. Their element flew by night into the small coastal town of Pasni, Pakistan. There they quickly boarded an AC-130 airplane and flew the rest of the way into Jacobabad. On the plane, Baker considered himself lucky to have such a great team.

His sniper team's camaraderie was hard to duplicate. The assistant team leader was Connelly, from Nebraska, and his laid back attitude reminded Baker of a carefree surfer's. Angelo, the team's radio operator, grew up boxing and fighting in Philadelphia. He had a "never quit" attitude. Lastly was Arne from Seattle. He loved to drink beer. Baker enjoyed that his marines played hard, but more importantly, that they worked hard, especially at sniping.

When their AC-130 landed, the marines went to work. Baker's element immediately established a perimeter. Grunts applied concertina wire and dug into defensive positions, while the snipers moved atop a building and set about finding ranges to target reference points. The commander informed

Baker, however, that he wanted the snipers primarily to observe the perimeter for people trying to infiltrate.

The next morning, Baker stood face-to-face with the reality before him. Sunlight exposed the nearby city while the scent of burning garbage lingered in the air. In a haze, aged mosques and dusty brick buildings gave Baker the feeling that he was in a third world country, and when the call to prayer resounded through the city, it sent chills up his spine.

He was roughly eight thousand miles from his hometown, Kalispell, Montana. Joining the Marines seemed natural after having grown up in the outdoors and witnessing how the Marines had shaped his brother, who had been assigned to Security Forces. Baker always enjoyed stories of those in the military, and after graduating high school early in 1998, he was shipped to boot camp.

In Pakistan, the snipers kept security at the airfield for a month. Being there was bittersweet. Baker was glad that he was doing something other than sitting on ship, but holding security was nothing like combat. He wanted to be on the other side of the border, in Afghanistan, where all the action was.

Soon other forces began to arrive at Jacobabad. Navy SEALs, Air Force personnel, and other members of the Fifteenth MEU slowly settled in. In that time Baker made friends with Marco, a Navy SEAL sniper. Typical of snipers, they talked about ballistics, range cards, tactics, and weapons. One day Marco stated that he was pushing forward into Afghanistan. He talked of a lone compound with guard towers. Only later did Baker find out that it was an airstrip that he and the others marines would help to seize.

Entering the Fight

Soon the marines from the Fifteenth MEU were directed to enter the war in Afghanistan. They fell under Task Force 58, with the mission to bring chaos to the remaining Taliban and al-Qaeda fighters in the south. They would establish the first American foothold in the country, known as Forward Operating Base Rhino, and from there would push out for follow-on missions. Their presence would mark the first conventional troops to enter the war. Until then, the only U.S. forces in the fight were derived from the United States Special Operations Command and other government agencies.

When Baker learned he was heading in, anticipation and exhilaration were his only thoughts. As a trained sniper, he wanted to use all that he had been taught. It took a few days to fully settle on the idea of actually going into combat, but on November 24, Baker and his team prepared for Afghanistan. The marines from the Fifteenth MEU, in keeping with the agreement between the U.S. and Pakistan, were flown to their ships off the coast to avoid launching from within Pakistan.

Task Force 58 flew into Afghanistan on the night of November 25, 2001, to seize FOB Rhino. Baker and his men carried combat loads; they had two M40A1 sniper rifles, a SAW, an M16 with an M203 grenade launcher, and multiple radios to cover every aspect of communications, to include satellite communications. This was the big show, and they knew that anything could happen.

The journey covered 441 miles, and during the helicopter ride, Baker wondered what awaited them. He was informed that British Special Air Service had cleared that part of the country, and that little activity should be expected, but still it was Taliban territory, and the enemy could be waiting to ambush his group. Days earlier, however, a team of Navy SEALs had infiltrated the region to provide reconnaissance and surveillance for the marines. They, along with Air Force Combat Controllers, had secured a landing zone before the task force arrived.

The CH-53s from the Fifteenth MEU swooped in hard and fast. When Baker's bird touched down, the marines flooded off the back, immediately forming security around the battalion commander. A land of isolation met the marines, and all that Baker noticed was dirt and sand—and the cold.

That night, Baker and the others held a perimeter, allowing more troops to move in. The next day, engineers arrived to clear the nearby buildings of booby traps and mines. When they finished, Baker's team moved onto a rooftop. Within twenty-four hours, Task Force 58 had cleared and occupied Forward Operating Base Rhino; it was a statement to terrorists in the country that conventional U.S. troops were moving in and were ready to fight.

Not long afterward, Baker's team received their first combat mission. This being their first time operating in such an environment, the snipers were nervous. Their mission was relatively simple. They were sent out to provide early warning detection for the FOB. Baker had two concerns. First was about navigating the open desert with no significant terrain features to base their movement on. His biggest concern, however, was the ever dangerous and hidden land mines. Over and over, the marines were warned about the hazard of land mines and that Afghanistan was riddled with them.

Before the snipers started their mission, Baker shot an azimuth with his compass, and the snipers patrolled on foot from the base. While they walked, Baker scanned his sector but cringed with every step. Everything they had been taught about patrolling was put to use, and after a short while, the snipers were in position.

They set up on a hillside, but a sandy bluff stood between them and the road they were to watch. Baker and Angelo, the radio operator, cleared the bluff and began to dig in. Later, they heard more digging from nearby and radioed the battalion, asking if anybody else was out there, and found that other marines were nearby. That night, the only event that happened was that Navy SEALs sped by on the road with infrared strobe lights flashing from their vehicles.

The next missions were much the same. As forces built up at FOB Rhino, the snipers were given vehicles to operate with. Baker's team was issued a Humvee to move out farther for early warning detection. Besides meaning they didn't have to walk, the vehicle let them carry a week's worth of chow, water, and ammo. For concealment, they positioned the Humvee in a crevice with camouflaged netting covering it. That week, the only thing they saw was camels with bells attached to their necks. Baker eventually realized that very few people lived in the Registan Desert and that they were not likely to find enemy fighters. The mission felt almost like a combined arms exercise in Twenty-nine Palms, California, but when they returned to base, Baker learned that his team was headed north.

The snipers were directed to provide more early warning detection for a forward patrol base. They would be closer to Kandahar, a major city, and the possibility of action was better. That information motivated the snipers. They did not want their entire stay in Afghanistan to be dull. The snipers gathered their equipment and one night boarded CH-53s for a forty-five-minute ride north, before dismounting and setting up a small patrol base. It was more of holding security for Baker's men, but they quickly discovered that they were needed to support an ambush on Route 1, a road leading into

Kandahar. The intent was to stop or capture enemy fighters moving from the city.

One Shot One Kill

The anticipation of finally meeting the enemy kept Baker awake on the night of the mission. He and Arne rode inside a Light Armored Vehicle (LAV). Most of the MSPF team were in their element, and they had practiced for just such ambushes in training. The marines cut through the desert night for hours before reaching their destination around 0300 hours.

Baker peered through the hull of his vehicle at the stars outside. More than anything, he was glad to finally be stopped. Outside, other marines began to form a blocking position on the two-lane road. Scouts from the LAVs dismounted and ran concertina wire across the pavement attaching chem-lights to the wire to warn traffic. Force Recon marines hid nearby, and when civilian traffic stopped, they would be the ones to pull the drivers and passengers from their vehicles. Baker would be covering them behind the rifle while the other marines held their distance.

A few hundred meters from the road, Baker set up his shooting position. He rested his sniper rifle on his pack atop the LAV. The PVS-10 day/night sniper scope provided perfect sight, and he and Arne discussed the distance to the concertina wire. They also talked about how they would react if the marines made contact. Overhead, aircraft surveyed the road from Kandahar and warned the marines of an approaching convoy. Baker heard the radio transmission, and shortly thereafter, he saw headlights on the horizon.

"We got a breeze from right to left, two miles per hour," said Arne. It was not strong enough to change Baker's windage.

"What do you think is going to happen when the vehicles hit the wire?" Baker pondered.

As the convoy approached, all the vehicles stopped in the distance except one truck. It proceeded to fly toward the concertina wire. Its headlights lit up Baker's night vision, giving him a better view. The truck did not slow and drove over the wire, which became mangled on its undercarriage, forcing it to stop. Seconds passed before a man stepped from the right side door to examine the damage, but in a flash he climbed back in after noticing the Force Recon marines pulling onto the road behind them.

As the Force Recon marines drew down on the men, Baker grew tense. He saw the marines yelling at two men in the bed of the truck who were covered in blankets. Suddenly the men jumped to their feet holding AKs, but the marines beat them to the trigger and opened fire.

Immediately Baker prepared to shoot. Everything about his shooting position, including his bone support to his snug butt stock, was correct. He aimed at the left side of the windshield and calmly took a breath. He realized that this could be his one chance and it was now or never. The light was bright enough for him to see the windshield but not the men in the cab. He estimated their position before firing.

The first shot, a cold-bore shot, might have been a few inches off at worst. After firing, it seemed that his first bullet did not get a reaction from the truck. The marines inside the LAV, however, yelled for him to cease fire, scared that his round would strike one of the other marines, but Baker knew the trajectory. He had trained with the Force Recon marines and had made shots like that before.

Soon the truck was on fire. Bullets hit the gas tank, setting the vehicle ablaze and causing RPGs (Rocket-Propelled Grenades) to cook off. When it was fully engulfed, the man who had gotten out of the truck earlier exited again. Baker saw that his head and right shoulder were on fire. The man was still a threat, and Baker repositioned his crosshairs between the man's shoulder blades. He had already chambered another round and lightly squeezed the trigger.

The man fell instantly. He died on the road. Minutes later, the Force Recon marines moved forward and swept the area. When the smoke cleared, the marines had killed seven Taliban fighters, but their comrades drove off, only to be met with close air support. In the end more than twenty Taliban fighters were killed.

The ride back to base was long for Baker. For seventeen

hours, he thought about his first kill and the events that had happened. Though it was one engagement, it made Baker's time in Afghanistan worth it. When they arrived at the forward operating base, the marines debriefed the situation. Baker explained his story, and it was speculated that with his first shot he had killed the driver. That was why he never exited the vehicle. Around camp, a marine from his LAV tossed Baker the two empty casings that he had used. Baker kept them as a memento.

When the marines departed Afghanistan, Baker was satisfied with his time there. His role as a sniper covered surveillance, reconnaissance, and an engagement. He was one of the first Marine snipers to set foot in the country and to dispatch an enemy fighter. It was just the start of the war for conventional snipers, and for Baker, it was a success.

FOUR

SPEC-OPS

IN the covert world of Special Operations, sniping is a highly regarded skill. Though it is one among many, snipers within Special Operations Forces know they will be called upon more often than not, especially when precision is needed. However, sniping is more than just shooting, and Spec-Ops warriors understand this better than any. In Afghanistan, two very different operators, a Navy SEAL and a Green Beret, experienced sniping and all its facets.

Into the Blue

Chris, a Navy SEAL, wanted sniper training. He could have chosen the Naval Special Warfare sniper school, but his friend, an instructor at the First Marine Division Scout/Sniper School, guaranteed him a slot if he chose to go there. When he asked his platoon chief, the salty SEAL advised against it.

"You don't want to do that. Do you realize that you are going to go up there, throw a rucksack on, and get beat? You're going to be treated like dirt for ten weeks. Why would you want to do that?" asked his chief. He had been around long enough to know the reputation of U.S. Marines.

"I just want to do it. I've always wanted to be a sniper, and I want to go to that school," explained Chris.

After a second, his chief replied, "OK. You're an idiot, but go for it."

Chris could handle the games. After all, he had been a marine before. He had started his career as a Marine artilleryman and by his fourth year in, he had passed the reconnaissance screening, but because he was a 0811, field artillery cannoneer, he was denied entry into the special unit. It was not until he took his wife to the Hotel Del Coronado in San Diego for their anniversary and saw the "team guys" running by on the beach that he finally decided to pursue his childhood dream. In 1996 Chris attended Naval Special Warfare's Basic Underwater Demolition/SEAL training and started his journey to become a Navy SEAL.

Four years later, he was back with the Marines, but this time it was different. He was one of two SEALs at the Marine sniper school. His class, 04-01, was the last one of the fiscal year in 2001. It started in July, and for three months, Chris and the other SEALs played the games at sniper school, but did everything the SEAL way. During field week, while the marines drew pictures and field sketches of the objectives, Chris and his partner took pictures with high-powered cameras. While the marines loaded their packs with their heavy radios and other equipment, Chris used his special lightweight and compact issued gear that the Marines had not even heard of.

By September, Chris had passed all the qualifications. He was two days from graduation, when on the morning of the 11th he pulled into the sniper school parking lot and an instructor stopped him, asking if he'd heard the news.

"No, what's up?" replied Chris.

"We've just been attacked by terrorists," said the marine.

Chris strolled into the classroom to find the other students huddled around a radio. They listened in detail to exactly what had happened. The instructors handed out still photos of the planes crashing into the World Trade Center towers.

The SEALs knew what to expect, and within hours there was a phone call. It was for Chris, and when he answered, the voice on the other end was serious and strict.

"Petty Officer Osman, your platoon has been recalled."

This was not a drill, and the SEALs needed to report back to their team. Chris packed his gear. The marines asked him where he was going, but he could not talk about it. He said a final farewell and was gone.

The ride to Coronado was full of anticipation. Chris wondered if this could finally be the call he had always dreamed about—the call to go to war. At his base, guards searched everyone, and a long line congested the road to get in. They questioned Chris about the weapons in his truck, but when they found out he was a SEAL, the guards let him in.

In his team room, Chris learned his fate. The terrorist attacks called for swift retaliation, and there would be blood. His unit, SEAL Team Three, Echo Platoon, was a desert warfare team with an area of responsibility of Southwest Asia, and they were chosen to deploy immediately. His platoon began packing and when Chris was finished with the meeting, he packed as well. Each of them had a locker the size of a small room, full of gear and equipment. Chris pulled everything from flashlights to his ghillie suit and stuffed it into his bags.

As the platoon ordnance rep, Chris accounted for all of the platoon's weapons. The armory held their MK-23 and P226 Sig Sauer pistols, M4s with accessory kits, M60s, and M203 machine guns. These, along with M14 and MK-11 sniper rifles, were a few of their weapons. Altogether, it was a mini-arsenal but standard for a sixteen-man SEAL team. It took two days to pack everything, and when they were done the team had seven pallets of gear, all vital equipment necessary for a SEAL platoon deployment.

During that time, the team also underwent isolation. No phone calls, no leaving base. They could not even go to chow without being escorted. They also began paperwork and were ordered to fill out their last will and testament. Chris left everything to his wife and daughter.

By week's end, Chris was allowed home. He kissed his wife and daughter good-bye but could not tell them where he was going because he did not even know. Though it was a painful moment, the possibility of having to do it had always been in the back of his mind. The only difference was that now it was all real.

When they returned to base, the SEALs loaded up. A massive C5 cargo plane awaited them at North Island Naval Air Station in Coronado. The small group of warriors boarded the plane, and though Chris still had no clue as to where they were headed, he knew that they were going to fight terrorists, and that was all that mattered.

A Long Road

On September 11, 2001, a few thousand miles from Camp Pendleton, California, Bobby, an Army Special Forces (SF) soldier, prepared for combat divers course. At Fort Bragg, North Carolina, he arrived at the pool early to practice Navy drown-proofing.

At the deep end of the pool, straps were bound to his wrists and ankles. The idea was to push off the bottom and get a breath at the top without using his hands or feet for propulsion. Half-way through the exercise, when he lifted his head from the water for a breath of air, Bobby heard the female lifeguard crying.

"It's so horrible," she said.

Underwater, Bobby wondered what she was talking about. On his next grasp for air, the woman screamed again.

"I can't believe they hit the buildings! Everyone's dead!"

"What on earth is she talking about?" thought Bobby.

His next time up, a friend grabbed him by the arm.

"There's some bad stuff going on. You need to see this," he said.

Bobby got out of the pool and walked into the lifeguard's office dripping wet. He watched the planes crashing on TV, and after a few minutes, he and his friend went back to their team room to learn the entirety of the attacks.

Bobby was in a U.S. Army Special Forces Operational Detachment Alpha (ODA) unit, also known as an A-team. As part of Third Special Forces Group, his team could deploy at any moment and were capable of facing just about any contingency thrown at them. The bad news, however, was that after al-Qaeda claimed responsibility, United States Special Operations Command (USSOCOM) formed a plan that kept Bobby's unit out of the first few months of the War on Terror. He would have to wait to see action. Just like any soldier, he did not want to wait. He had done plenty of waiting, and he wanted to finally put to use the years and years of training.

Bobby had started his Army career in the regular infantry. In the Eighty-second Airborne Division, he had done almost every job available to an infantryman. He was a point man, breacher, Squad Automatic Weapons (SAW) gunner, radio operator, and machine gunner, but deep down, he wanted more. By the time he was an E4 specialist, two opportunities came his way to do just that.

He had always wanted to be an Army Ranger. When he was twelve years old, his family went to an air show on a local Army base. Bobby watched in awe as Rangers jumped from airplanes, parachuting onto the airstrip in front of him. He noticed everything about them—their camouflaged face paint and weapons and their beaming confidence. When the Rangers reached the crowd, they gave away recruiting flyers, but Bobby did not take one. He already knew that he was joining the Army, and he wanted to be a Ranger.

While still an infantryman, Bobby begged for the chance to go to Ranger school. After he discovered that one of the squad leaders in his platoon had failed the physical training test on the first day of Ranger school, Bobby walked straight into his platoon sergeant's office.

"Sergeant, I think I'm ready to go to Ranger school," he said.

"What makes you so sure that you're ready to go?" inquired his platoon sergeant.

"Well," he said, "I guarantee you that I won't go down there and fail the PT test."

Bobby's confidence won over his leader—that and his stellar physical fitness performance. He consistently ran a sub twelve-minute two-mile. His platoon sergeant directed him to the soldier in charge of allocating schools, and before he knew it, Bobby got his wish.

In the summer of 1996, Specialist Bobby checked into Ranger school. The beginning test, the Ranger assessment phase, indicated how tough the school would be. Push-ups, sit-ups, pull-ups, a five-mile run under forty minutes, a water survival assessment, a day-and-night land navigation course with a map and compass, and a twelve-mile hump with full gear were just the start. It was a grinder, but the school only got harder.

For sixty-one days straight, Bobby experienced the nightmare that is Ranger school. Of everything he confronted— food and sleep deprivation, extreme physical and mental fatigue, rugged mountains and exhausting swamps—his time with one soldier left the most memorable impression.

His first Ranger buddy was injured early on. Ranger Dobb, his second Ranger buddy, was an Eighteen Delta, Special Forces medic. Behind the legions of patches, and the lore of Special Forces, Dobb was a humble, down-to-earth soldier. Dobb was his first name, and when the class first started, he asked the other students to call him by it. Allowing those of lower rank to address him in that way is not normal for a staff sergeant. He also gave his fellow students advice.

"If you guys have any medical concerns, get with me first. I'll try and help you as best I can," Dobb said. In Ranger school, students hide injuries to avoid being set back in the training.

Unfortunately, professional jealousy plagued Dobb. No soldier wants to believe that another unit is better than theirs. Throughout Ranger school, Bobby noticed animosity toward his partner, Ranger Dobb, from the Ranger instructors, or RIs. It amused Bobby that though the instructors tried harassing him, Ranger Dobb always came out on top. Once a fellow student requested Dobb's help over the RI medic, causing an instructor to berate Dobb in front of the entire class.

High in the mountains, the students were in the middle of a long hike. One soldier slipped from the trail and rolled down the mountainside. Everyone heard him tumble, and when he stopped, the student let out a painful scream.

"Dobb!" he cried, sending an echo throughout the valley.

"Who the hell is Dobb?" yelled an RI.

"Doooobb!" again called the student.

By now the RI was fuming.

"I wanna know. Who the hell is Dobb?" he yelled.

"I am," said Dobb. The RIs only knew him by his last name.

"What's so special about you?" said the instructor.

"Well, I'm an Eighteen Delta," he humbly replied. "Do you want me to go down and check him out?"

At that, the instructor lost it.

"Are you kidding me? You're no better than our medics!" he yelled. "You're just a student. You think you're special because you're Special Forces!"

In the end, however, Dobb had the last laugh. He was the class's distinguished honor graduate. Dobb's poise under pressure and his incredible leadership never left Bobby. He had never met a Green Beret before, but while working with Dobb, he watched his every move and learned what it meant to be a professional soldier. Dobb helped to propel him to Special Forces.

Weeks after the terrorist attacks, Bobby's team wished the men from Fifth Group luck. They were headed to Afghanistan. Bobby knew he would get his time, but for now, it was back to training. A few months later, Bobby was in Florida with the rest of his team conducting refresher training for close air support. Their training was cut short, however, and his A-team was recalled to deploy to Afghanistan.

Sea to Land

SEALs are trained for all environments. They cover sea, air, and land as their name explains, which makes them versatile for different missions. They are capable of foreign internal defense, information warfare, security assistance, direct action, unconventional warfare, and more. However, underneath it all, SEALs are the best at maritime operations. Their skill in the water and everything associated with it separates them from other Special Operations units.

It made sense to Chris that when they landed in Kuwait, his platoon would be doing ship takedowns in the Persian Gulf. Visit, Board, Search, and Seizure, or VBSS as it is known, was their bread and butter. Chris's platoon arrived to relieve the SEALs already there. The other SEAL platoon was moving forward to Afghanistan. When Chris's platoon heard that, their enthusiasm about doing ship takedowns was almost lost; normally the SEALs did not mind doing them, but since there was a war, they wanted to be in it.

The platoon made the best of their situation. They began night patrols along the coast to interdict hostile vessels. Their boat, an MK V Special Operations Craft, held the entire team and was able to speed at fifty-four miles per hour. As a platoon sniper, Chris normally used his M4 and rotated with other qualified snipers in his platoon for the duty of sniper over-watch. If called upon, Chris would hold sniper over-watch from a helicopter while the rest of his platoon boarded the ship. After three weeks, they had yet to see action, until one night they finally got the call.

The SEALs were hustled into their ship's cabin. Chris stood in the back listening to the situation.

"There's a ship that needs to be taken down," explained their unit commander. "Its name is *Alpha 117*. This ship was used to smuggle explosives into Africa. It destroyed multiple U.S. embassies, killing two hundred and injuring a few thousand people. You're taking the lead, so jock-up. You guys are going right now!"

The SEALs rushed into action. Chris would not be a sniper on this mission. Instead, he would board the ship with the others. After suiting in essential gear and double-checking their weapons, the SEALs were ready in minutes. Soon they moved onto a smaller vessel, their Rigid Inflatable Boat (RIB), and sped toward the target.

It was pitch black when their RIB closed in on the target ship, already under way. When they pulled alongside it, the pole man hooked the railing, and the SEALs scurried up the ladder onto the deck.

Atop, Chris and six others waited for the rest of the men to climb up. They looked frightening; they all wore black balaclavas to cover their heads and woodland battle dress uniforms, while gripping M4 assault rifles. On their knees, the SEALs held security, but suddenly the cabin door swung open.

"Open door!" yelled one of the SEALs, and the team flooded toward it.

The man opening the door was smashed in the face and knocked over. Chris pulled him out of the way to let the other SEALs rush inside. He quickly zip-tied the prisoner's hands and feet, and when the man began struggling, another SEAL,

Chris's cover man, bounced his head against the deck, yelling for him to "Stay down."

Chris was assigned to the engine room team. He passed the prisoner off and fell in with three other team members. Their movement through the ship was fluid. They had trained a thousand times for ship takedowns, and now that it was real, they were well prepared. The heat from the ship had Chris sweating profusely, and when they entered the engine room, it was even worse.

Three men were working inside the engine room. When the SEALs ran in, the men lifted their hands and surrendered. Within three minutes, the ship was under control of the SEALs, and after they moved the ship's crew to the back deck, a thorough search of the entire ship was done. Though no shots were fired, Chris enjoyed the mission more than any others. This was a true SEAL mission and at the time applicable only to the men of Naval Special Warfare. He even enjoyed the mission over his time in Afghanistan.

By late 2001, Chris's team had been re-tasked. They were flown into Oman, a country on the Arabian Peninsula. Naval Special Warfare would be involved with a special unit hunting and tracking high-value targets in Afghanistan. SEAL Team Three, Echo Platoon, would be part of the group, and Chris's platoon was headed to the desert.

With their new land warfare mission, the SEALs began

preparing. Painted weapons and equipment lay in their berthing area, and at that time, most of the SEALs had not used Humvees before. Chris, having been a Marine, was very familiar with them. With the help of another platoon from SEAL Team Three and a platoon from SEAL Team Eight, he helped to train his platoon on night driving, basic mechanical skills, and immediate action drills. He also pulled out his sniper rifles. He had a .300 Winchester Magnum and an M14 with match-grade ammo. Should he need them, they would be good tools for the distances of Afghanistan.

Task Force K-bar was their new assignment. This elite group was formed with Special Operations units from seven different countries. The Task Force was given the mission to destroy, degrade, and neutralize the leadership of al-Qaeda and the Taliban. They learned of terrorist training camps, bomb-making factories, civilian and enemy personnel, and where to find al-Qaeda and the Taliban. They were to be on the hunt the entire time. Chris's platoon, in particular, would be hunting in the eastern region of the country. There they expected difficult terrain and cold weather.

Echo Platoon touched down at Kandahar Airport in late December 2001. Right away the cold was felt. Three hours before, the temperature had been over 120°F; now it was just above 60°F. The SEALs were given no time to acclimate. The enemy needed to be hunted, and the SEALs needed to get into action.

Before Chris's arrival, the Taliban had been removed from power. The Northern Alliance and their allies, U.S. Special Forces, and CIA operatives had pushed the Taliban from the major cities, while al-Qaeda had been driven to the countryside. However, a few hours north of Kandahar, in the Paktika Province, al-Qaeda had regrouped near the town of Khost.

That group occupied the Zhawar Kili cave complex. This complex was an enormous expanse of underground tunnels, with compounds and buildings nearby. Its use was for holding supplies and terrorist training camps, and for storing ammunition and troops. Al-Qaeda leadership also planned their missions there. Initially, an Army Special Forces Operational Detachment Alpha (ODA) team was tasked with the mission, but certain events led to Chris's platoon from SEAL Team Three taking over. They, along with a few attachments, were dispatched to clear the caves.

FID/Recon

Bobby was excited about Afghanistan, but his first deployment there saw limited sniper operations. Before leaving Fort Bragg, his team learned that they would be doing FID, or Foreign Internal Defense, training. Essentially, they would teach the Afghanistan National Army how to fight; it was just the type of mission especially suited for Special Forces soldiers.

FID is one of many missions for an Operational Detachment Alpha. A-teams are comprised of twelve men, expertly trained at infiltrating remote, hostile countries, finding friendly local-nationals, and training them on weapons and tactics. Each member learns this during five extremely difficult training phases known as the Q-course. Those who pass earn the eighteen-series military occupational specialties and the honor to be called a Green Beret. For Bobby, the training had happened years ago, and now he would be putting it to use.

His A-team arrived in Afghanistan's capital city, Kabul. Right away, they met the indigenous troops they would be training. The job was tricky business for the SF soldiers. Bobby was honored to train the very first troops of the new Afghan government, but there was a harsh reality behind it. The possibility of al-Qaeda and Taliban insurgents infiltrating the ranks to learn the Army's capabilities and strategies was highly likely, but the job had to be done.

Bobby was one of the two 18Bs, weapons sergeants, in his team. His expertise dealt with the employment of several types of weapons both foreign and domestic. They included small arms, light and heavy crew-served weapons, and antiaircraft and antiarmor weapons. He had also become proficient with sniper rifles before joining SF.

In the Eighty-second Airborne Division, Bobby had been in a scout platoon. There, he was given the chance at scout/ sniper school at Fort Benning, Georgia. The elite of the infantry, the scouts made up the reconnaissance and sniping element for the battalion. Bobby's conclusion about sniping was simple. Just as with the animals he had killed for food growing up, if he had to, he would snipe another human and with no problems.

From Ranger school to sniper school, Bobby was bound for another dreadful adventure. The students stayed in a barracks built in World War II. Within seventy-two hours of arriving, Bobby and the others were forced to make ghillie suits from scratch. From there, students dropped from the course like flies after failing stalking and shooting. Bobby was no stranger to scoped weapons. He had hunted with them since childhood in the foothills of the Blue Ridge Mountains. He enjoyed the challenge of shooting targets at extended ranges, even when he was forced to find the distances during the unknown distance phase of the training.

His most memorable experience, though, was the FTX, or the Final Training Exercise. For a week straight, the students were forced to use everything they had learned. Planning, patrolling, stalking, reporting, and the final exercise: one shot at an unknown distance target. After six long weeks, Bobby graduated sniper school and received the B4, sniper identifier military occupational specialty.

In Kabul, Bobby's team trained Afghan troops for three months. He'd taken six months of Arab language instruction but spoke none of the tribal languages of Afghanistan and needed an interpreter to teach patrolling, ambushes, and ways to use terrain and heavy weapons to advantage. Teaching indigenous troops is fun, but not as satisfying as hunting enemy troops, and Bobby was relieved to hear that after the third month, his team would be transferred to do just that.

From Kabul, Bobby's team convoyed south to Kandahar, the last stronghold of the Taliban among Afghanistan's thirty-four provinces. Special Operations Forces were sent to track the remaining leaders. Bobby's team arrived to be part of a joint task force operation, a mission to apprehend high-value targets.

The nearby town of Deh Rawod supported the Taliban. Mullah Mohammed Omar, the supreme leader of the Taliban, and other Taliban leaders had family in the area and trusted the town as a sanctuary. U.S. Special Operations, foreign coalition forces, and Afghan troops would search the town for the leaders or any other Taliban militants. For the operation, Bobby's team was to conduct a visual recon of the town in advance of the main force and, if need be, to guide bombs onto targets. Bobby, in particular, acted as the sniper in his team, and though he didn't engage, the operation was the perfect example of how snipers perform.

On subsequent nights, Bobby, his team, two Air Force tactical air controllers, and two communications soldiers (to intercept radio traffic) infiltrated the mountains around the town. On their way in, the helicopters took fire from Anti-Aircraft

Artillery (AAA) weapons inside Deh Rawod, but the team landed safely at around eleven thousand feet elevation. Immediately one of the communications soldiers showed signs of altitude sickness. He was dizzy, had a headache, and couldn't walk more than a few hundred feet before becoming completely fatigued. Bobby had an M24 sniper rifle and a week's worth of equipment on his back, as did the other soldiers, but they took turns carrying the sick soldier's hundred-pound pack, making the already tough three-kilometer (two-mile) movement even more grueling.

The men were in position before daybreak. Bobby helped to blend the team's location into the mountainside. Their hide encompassed everyone and kept them all unseen, and Bobby easily observed the town, five kilometers away. While watching the area, they noticed that as friendly aircraft flew nearby, antiaircraft artillery came to life from inside the town and attempted to down the planes. On the night when ground forces proceeded to insert, AAA weapons became the main threat as the troops traveled into the town by helicopters. When Bobby's team reported the activity, the word came back to direct air support onto the positions, giving the tactical air controllers free rein.

The compounds were leveled before Bobby's eyes. When the dust cleared, the damage was irreparable. Taliban fighters commonly hid behind civilians, but Bobby was shocked to find that the fighters had strategically placed the AAA weapons close to the civilian population in hopes that they would not be targeted. As a result, civilians were hit and lives were lost. It was not the type of outcome Special Forces soldiers enjoyed, but though the incident was tragic, after the bombings AAA weapons did not fire at Coalition Forces again.

Ever ready at a moment's notice, Bobby and his team traveled north to Asadabad. They moved in specially configured Humvees known as Ground Mobility Vehicles (GMVs), fitted with little to no armor, to keep the vehicles faster. The treacherous roads made it hard for the normal, armored Humvees to navigate, and SF soldiers didn't trust them to go off-road into sand and other difficult terrain. The GMVs didn't have doors and had three mounted weapons, one .50-cal and two M240 machine guns, bringing much more firepower to the fight than the normal vehicles.

While patrolling near Asadabad, Bobby survived his first IED attack. He was manning the .50-cal on the road, when right beside them on the road an IED exploded. In the explosion, shrapnel peppered the entire vehicle and dust covered the road. Bobby's driver punched the gas and sent them off the edge of a ravine and down the side of a hill. Miraculously nobody was hit, but a passenger broke his arm in the crash.

Bobby also learned the differences in cultural customs there. The locals still lived in the Stone Age. They had never seen the weapons, gear, or equipment the U.S. military had, especially SF. Also, the treatment of women and children by the men was severe. Once, Bobby's team patrolled through a small town, and while driving, Bobby waved at a little girl, who smiled back with a wave. When her father saw that, he picked up a stone and hit her in the head with it, knocking her down. Bobby reacted on instinct; he kicked the driver and told him to stop. Before he climbed out of the GMV, he grabbed his M4, then stormed up to the man, barrel punching him in the chest.

"Tell this man, if he ever does that again, we'll leave him in the desert to die!" he yelled to the interpreter. Bobby thought about his own daughter and could not imagine doing anything like that to her.

Another time, he witnessed the strength of the kids. His team was at a remote outpost, in very bad weather. The wind blew so hard that their fire flickered sidewise. Bobby was in the best cold weather gear available on the market. He was wrapped in sweaters, thermals, hiking boots, and jackets, but he was still cold and moved closer to the flame. When he looked beside him, a six-year-old boy in a tattered sweatshirt, sweatpants, and rubber boots was amused that the American soldiers were cold, because he was not in the slightest bit.

At the end of seven months, Bobby returned to Fort Bragg. It was a relief to get back to his wife and to be in the States again. A month later, however, he was back to training, and went through the Special Operations Target Interdiction Course. It was a gentleman's course, considering that the students were not physically destroyed by the instructors. They didn't need to be, though, because most of the students had already passed the Q-course. By the time they entered SOTIC, they had proven themselves physically, and the instructors wasted no time smoking the soldiers. Instead, they focused on training, and that was how most SF schools happened.

Cave Clearing

The Zhawar Kili caves were frightening. In January 2002, Chris's platoon; two Navy explosive ordnance disposal techs; two Air Force combat controllers; an Army nuclear, biological, and chemical warfare soldier; fifty marines; and two FBI agents were set to move in. The group was informed that bombs had destroyed everything and that they would be assessing battle damage on the area. DNA samples, fingerprints, fingernails, and hair samples were to be collected to see if any high-value targets had been killed. It was likely that al-Qaeda was still in the region, and possibly even Osama bin Laden was still there.

The flight to the area took three hours. The teams were shuttled in by CH-53 Sea Stallions and dropped seven miles from the objective. Sparse vegetation sprinkled the barren landscape, and it was amazing that anyone wanted to live in such a remote location. The world's most sought after terrorists, however, had no choice. This was their backyard.

Within the group, each unit had separate objectives. The marines would hold security for everyone, while the SEALs were to do the cave clearing. The explosive ordnance disposal sailors would destroy any explosives found, the combat controllers would direct air support, and the FBI agents would examine the dead. When they finally reached the area, they realized that the bombs had done little to no damage to the tunnels and infrastructures.

The caves were dug into a cliff on the side of a small riverbed. Above the cliff was a small village. Other small buildings were in the riverbed. As the team closed in, Chris noticed that the cave entrances were semi-buried. The bombs had missed their targets and had thrown dirt everywhere. The combat controllers marked the smaller entrances with a GPS, and when they reached the main tunnels, the SEALs prepared to enter.

Searching the caves was nerve-wracking. It would be close-quarters combat if anyone inside decided to fight. Chris anticipated everything from enemy fighters to land mines. To add to the stress, when he started in, the light from his surefire flashlight disappeared in the blackness.

The tunnels were big enough to drive a truck through. The SEALs pushed inside while the marines held the high ground outside. Chris stuck close to the man in front of him but could not see much in the darkness. He prayed that they would not step on any booby traps. As they made their way deeper, the SEALs began to discover random items.

"Look," said Chris's teammate, motioning to the ground.

Chris glanced down at a human foot in the dirt. Moments later, they came upon a jail cell, complete with bars and locks. Next, they found small rooms. Chris entered while others held security. His light revealed boxes upon boxes, stacked from the floor to the ceiling, full of small arms ammunition.

Throughout the caves, the SEALs found more weapons and explosives, enough to supply a small army. Surface-to-air missiles were uncovered along with tanks and Russian-made amphibious track vehicles known as BMPs. They found rifles, land mines, mortars, and radios. By the end of the clearings, Chris was amazed at the inventory. They found literally hundreds of thousands of rounds of small arms ammunition, thousands of rifles, tens of thousands of mortars and land mines, but there were no bodies or men. The occupiers must have fled into Pakistan, only three kilometers (two miles) away.

After hours in the caves, the group prepared to patrol back to their pickup point. Before leaving, they marked the locations of the cave entrances and the vehicles for follow-on bombing. All the while, the FBI agents attached had not found bodies to examine. It was no surprise to them, knowing that in the enemy's culture, burials happened within twenty-four hours of death. The problem was that when the commander learned that no dead had been found, he canceled the extract ten minutes before pickup and ordered the group back for further searching.

The group decided to hole up in the village built on the hill above them. Chris should have known that nothing in war ever happens as planned. He had packed for a twelve-hour mission, but now there was no telling how long it could take.

Under darkness, the men patrolled back. When they arrived, Chris and two other SEALs were sent to search the buildings inside the village. The area was empty and the group took one of the buildings. Inside, a small cooking stove was used for warmth as nobody had brought any sleeping materials. Within a few minutes the men were fast asleep.

While they slept, AC-130 Spectre gunships patrolled overhead. Their thermal imagery spotted militants nearby and killed them. Within hours, the men on the ground were directed to search the remains of the dead. Before sunrise, the group split into two elements. Chris would be with the group searching the dead a few kilometers away.

They arrived before daybreak at the spot of the bombing. The AC-130 retreated to base, no longer having night to cover it. Another reconnaissance aircraft guided the team onto the location. When they neared the position the AC-130 had bombed, their radio cracked up.

"You guys should be standing on the bodies," said the radio operator on the aircraft, but there were no dead in sight.

Everyone stopped and took a knee. Chris and the others set security while the platoon commander began to scout the area. Within seconds, voices sounded behind Chris's team.

"Go to ground," whispered the platoon commander, immediately sending the SEALs prone and hugging the dirt.

Chris searched for the voices. They came from a group of men he thought were close to 1,000 meters (3,300 feet) away and climbing from a cave with weapons.

"Hey, man, how far do you think they are?" the platoon commander asked Chris.

"About a thousand meters," Chris replied, but he misjudged the distance. With the rolling hills, it was hard to determine the range.

"Call in an air strike," the officer said to a combat controller.

Minutes later, F-18s flew in hot. The whistle of falling bombs had everyone in the team watching. They all witnessed the bombs miss their mark low and to the left.

The next bird on station was a B-52. It was close to dropping its ordnance, but the pilot wanted the SEALs' location first. The last thing the SEALs wanted was to give their coordinates to the bomber, especially when months earlier Army SF soldiers were killed after being accidentally targeted in a similar scenario.

The team did not give their coordinates and the planes would not drop bombs. It took the authorization from the platoon commander and his stating that he would take responsibility for the placement of the ordnance for the pilots to send their payload. Only after the officer stated his initials over the radio were the bombs released.

Meanwhile, the SEALs shot at the militants. At the distance, effects on targets were difficult to gauge. Chris had his M4 with a scope made for close-quarters combat and tried his best to apply Kentucky windage to get a hit. Surprisingly, the militants had no clue where the fire came from and did not run. They stayed very close to the cave entrance and began searching for the origin of the shooting.

"Bombs away. Impact forty-five seconds," said the radio operator on the B-52.

Chris braced for impact behind a small mound of dirt. When they hit, the fourteen Joint Direct Attack Munitions (JDAMs) sounded like popcorn raking the mountainside. One landed five hundred meters from the group. The ground rumbled underneath them. The bomb fell at such close proximity that rocks and pebbles landed around them. When the dust settled, half of the mountainside had collapsed upon itself, and trees and brush were ablaze.

"Let's go," said the officer in charge.

The men quickly headed to assess the damage. Less than a hundred yards away, they came upon a small house. When they noticed it, the men rushed to the entrance and cleared the building. Inside, jugs of water and wooden trunks loaded with passports, money, and freshly pressed clothes lay about. Since everyone had prepared for a twelve-hour mission, Chris's team had drunk most of their water. The jugs in the room were heaven-sent, allowing everyone to top off his canteen. As they left, they marked the house with a GPS and headed for the mountainside.

When they reached the craters, smoldering rubble was all that was left.

"Did we even hit these guys?" questioned Chris. There were no bodies anywhere.

As they walked up the hill, a small cutout room was left. Stacks of blankets, cooking wear, and rifles were inside. The SEALs broke the cooking wear, took the rifles, and moved on. A mound of rubble was all that remained of the main entrance to the cave where the men had been spotted. Black, smoldering trees around the area held shards of burned clothing. Unknown to the SEALs, a P-3 Orion had recorded the bombing and revealed twelve to fifteen men completely obliterated. After a quick search, the team patrolled back and rejoined the other group. As they left the area, they called in an air strike on the small house they had searched.

Their time in the area lasted a total of nine days. The SEALs took follow-on missions to search the surrounding area and nearby towns as well. On day three, they were resupplied, and

Chris's sniper rifle was flown in. A few days later, the men moved to clear a village close by, and there Chris and another SEAL provided sniper over-watch.

On the morning of the mission, while it was still dark, Chris and the other sniper set out for the village. They planned to watch the area in advance of the main effort. It was eerie patrolling in the hills, just the two of them. By daybreak, they were above the village, hidden among rocks. Their old maps did not give them an accurate picture of their initial position. It was a cliff face, and they moved to their secondary position.

Hours later, Chris watched the group search the village. He and his partner calculated several different ranges in case they had to engage, but there wasn't much action. The village was empty, and the main effort filed out of the area. Chris and his partner covered their movement and prepared to meet them, but before they left, Chris's man-tracking skills came into play. He noticed fresh footprints on a goat trail leading in the opposite direction. The two of them wanted to see where the tracks led and radioed their commander.

"We've got fresh tracks here," Chris said. "We're gonna follow them."

They received the OK and moved out. On the trail, Chris was worried. He carried only a bolt-action sniper rifle, and if they made contact, he'd be stuck with a single-shot weapon. The trail led them down a hill and directly into a cache of clothing and three sleeping bags. Clearly the owners of the materials had tried hiding them. Chris marked the position with his GPS and kept walking. Soon, the trail split into two, one aimed toward Pakistan, while the other led to another nearby village. The SEALs took the trail toward the village. As they patrolled, Chris's partner suddenly stopped him.

"Look," he said, pointing at a bunker.

Thankfully, nobody was manning it, or Chris would have been dead. The hidden bunker was big enough for five guys. Inside were a fire pit, logs, and cooking materials. Chris also marked the bunker with his GPS and recorded the coordinates to the village. When they were finished, the two snipers patrolled back and met up with the rest of their unit. That night, a surveillance plane monitored the village that Chris and his partner had stumbled upon. Radio traffic indicated that Taliban militants held meetings there. They arrived that evening in two trucks. The surveillance plane recorded as bombs hit the vehicles, chalking up more kills for the SEALs.

By the ninth day, the SEALs had achieved the remarkable. They had collected an enormous amount of intelligence, discovered more than seventy caves, and called in nonstop air strikes for days. Being among the first Americans in the region, the mission was surreal for Chris. Thinking of all the 9/11 coverage on TV and all the talk of payback, he realized he was actually in the place al-Qaeda terrorists called home and presumably the area where the hijackers were trained. Although he would have liked to have had more trigger time with the sniper rifle, he had acted as a sniper and made it out alive.

When the mission was complete, the group headed back to base. They had completed two months of ship takedowns and four months of ground warfare in Afghanistan, and now it was time to head home.

PSD/Direct Action

SF soldiers are perfect for PSD, or Personal Security Detail. Specialized training makes ODA teams exceptional at protecting high-profile dignitaries; however, most SF soldiers will tell you that these missions are not preferred, and for two reasons. One, the soldiers are almost always on the defense, and two, usually there is not much action. On his second tour to Afghanistan, Bobby's team was tasked with training the men who would be protecting the newly elected Afghanistan president, Hamid Karzai.

Bobby had graduated SOTIC months before and could now put the tactics to use. In Kabul, he and civilian snipers from DynCorp International, a security contracting company, trained indigenous troops on the role of marksmen during personal security. Their focus concentrated heavily on counter-sniper roles, specifically for protection.

Most times, when the president traveled, Bobby and his group moved into the intended area days in advance. He taught the indigenous troops to set up landing zones for the president and how to scout the area where he would be staying. Normally, the snipers would observe the area from rooftops for days, and then secure the area before the president arrived. There is much to consider during counter-sniper operations, especially for PSD. Searching for possible enemy shooting positions, ambush sites, routes and routines, and suspicious activity is what counter-snipings is about. Spending hours behind a scope, meticulously combing buildings and roads exhausted Bobby. Fortunately, he enjoyed the company of his teammates, and they passed the time talking about family or home.

No matter where he was, Bobby always missed his family. That emotion was kept in the back of his mind so he could concentrate on the task at hand. The sacrifice made by the wives and girlfriends of Special Operations soldiers was hard to evaluate. Considering that the typical Green Beret undergoes years of intense specialized training all over the world, demanding total concentration, Bobby felt lucky to have a wife who would allow him to do his job.

When Hamid Karzai arrived, Bobby's element kept security. The snipers were always on high alert during the president's events and until he departed. Fortunately, no attempts on the president's life happened while Bobby's team guarded him, and in three quick months, Bobby's team returned to Fort Bragg. Their tour was up.

Bobby enjoyed his third tour the most. In January of 2004, his team arrived in Afghanistan once again, but this time for direct action. For four months they would be actively hunting terrorists and high-value targets in their home environment. Their new commander was just what SF soldiers want in a leader, a go-getter more concerned about war fighting than politics. Commanders make quite a difference for SF soldiers; some are concerned about their careers more than fighting and are timid about letting A-teams loose. Luckily for Bobby his new commander wasn't that type of soldier.

In no time, the team was kicking down doors. They started a cycle of operations beginning with direct action, where they hunted high-value targets for two weeks. From there, their commander sent them on probing missions into areas considered Taliban strongholds. For the sniper operations, Bobby carried an SR-25 mounted with a universal night scope, the PVS-22. The SR-25 gave him the ability of semiautomatic fire, which he definitely needed during raids.

Speed is a key element for direct action. A-teams raid targets hard and fast, leaving no time for the enemy to prepare or react. When Bobby's team hit houses, as the sniper it was his job to provide over-watch. Since the team used helicopters to move in, Bobby used a quick method to get into position. As soon as the ramp was lowered from his helicopter, he sped off the bird on an ATV to the nearest high ground. There, he set up a tripod and observed the back of the target houses for anyone trying to flee. Sometimes, when the birds landed, everyone in the house stumbled outside to see the commotion. In that environment—nighttime, with crowds of men, women, and children, and having to quickly identify actual targets—Bobby was thankful for all his training. More than once he had put his crosshairs on people who looked to be hostile, but turned out actually to be noncombatants.

After months of direct action, the A-team was re-tasked. They were to move into Shinkay near the Qalat Valley in the Zabul Province. This southeastern province troubled Coalition Forces. Its vast landscape, spanning the Afghanistan and Pakistan border, left many routes open to Taliban and al-Qaeda insurgents moving into Afghanistan. OGA (Other Government Agency) signal intelligence intercepted cell phone traffic indicating the nearby mountain ranges as safe havens for fighters crossing the border. Meals, beds, arms, and ammunition were distributed there, and from the valleys of those mountains, fighters moved into other regions of the country depending on the need for personnel. Finding these routes proved difficult, simply because there were not enough men on the ground. To counter, Bobby's team was to establish a firebase in the region. Accompanied by Afghan troops, they would begin finding and fighting the insurgents.

On their way in, Bobby's unit immediately suffered

casualities, and he was lucky to still be alive. The unimproved roads were filled with rocks, sand, and dirt, making it impossible to tell if anything was buried there. The conditions also wore on the vehicles, but the SF soldiers' GMVs worked better than the Afghans' light 4x4 pickup trucks. They were more powerful and had a wider wheelbase.

As they drove along the rough road, an explosion ripped through the vehicle behind Bobby's, and shrapnel tore into two of the Afghan soldiers, killing them. The device that they had triggered was a pressure-detonated IED. The only reason it had not detonated underneath Bobby's vehicle was because of the GMVs' wider wheelbase, which distributed weight better than the light pickup trucks. The loss was hard on everyone; Bobby's team had trained and worked with the Afghans for months and had built a good relationship with them.

Forming strong ties with the locals brings success for SF teams. Regular Army soldiers do not often train, work, and live side by side with indigenous troops, but for A-teams, establishing these bonds helps them to win over local support. When Bobby's team reached the valley, they immediately set up a firebase. Normally they hired locals from nearby villages to cook for them, but since their outpost was so far from any population, an Afghan soldier took over the duty. He was well paid, but also was warned that if the team were to get sick, he would be held responsible.

Trusting all of the Afghans was not easy. The SF soldiers

had a habit of sleeping with a pistol under their pillows and their M4s within reach. In some units, they caught Afghan soldiers with ties to the Taliban, and statistically speaking, there were more.

After weeks in the valley, the soldiers and the insurgents were destined to meet. The SF commander sped up the process and formed a plan. Since they were unable to pinpoint the enemy's exact location, even though they knew they were there, the plan was to trap them. The commander directed Bobby's team, along with the Afghan troops, to fly deep into the valley and hold a blocking position on a ridgeline. Another SF team started at the valley entrance and pushed forward, forcing the insurgents to fight or run.

On the night of the mission, as usual, the helicopters took fire on their way in. Bobby had remembered to test fire his sniper rifle before leaving. He always carried it, and was able to employ it, but had yet to fire it. This time would be different though. When they reached their dropoff point, it was actually a steep ridgeline. The birds were forced to put them down just above the area they were supposed to be watching, but as they did, RPGs began to fire on them.

Insurgents used tactics to try downing the helicopters. They would not use small arms or machine guns, because they wanted to keep their positions concealed. Instead, they learned to recognize the flashes of static from the main rotors above the birds and shot RPGs toward it. Only when they hit a helicopter and caused it to crash did they send in foot patrols to finish off survivors. The pilots, though, were unfazed and hovered in midair while dropping just the tail ramp onto a ledge. When the ramp hit the dirt, the soldiers piled off.

Bobby was the third man off his helicopter. On the way out, the man in front of him went left. That was Bobby's cue to go right. Outside, he heard the soldier in front of him yell. Only later did he learn that an RPG had missed the man by inches. When everyone was off, they all climbed a nearby ridgeline and began to search for targets. Bobby found a small nook between two rocks and set up a shooting position using a tripod to rest his rifle upon. That night produced heavy fog, hampering Bobby's vision, and rendered the night sight almost useless. He would have to wait until daybreak to see anything.

Both the insurgents and soldiers were aware of each other's presence, but could not see each other. A firefight was brewing, and the upper hand would go to whoever spotted the other side first. In the darkness, the soldiers prepared for an attack. They spread out on the ridgeline so that more than one of them would not be injured by an RPG or mortar fire. On the opposite ridgeline and the valley floor, insurgents prepared as well. They carried AKs and RPGs to several different positions on the mountainside. That allowed them to shoot and move without staying in place for too long.

As the sky lit up, Bobby packed away his night vision. His group was hunkered down when someone noticed smoke from a small fire on the valley floor below. Two SF soldiers decided to take a force of ten Afghans to scout it out, but just as they left, an enemy machine gun opened up from the other side.

"Cover us," said one of the soldiers.

"I got you," said Bobby.

Soon Bobby found his first target. It was still dawn, but on an opposing ridge, a man lay on his stomach behind a rock, aiming toward the soldiers moving below. Bobby quickly used his Leica Viper laser range finder and discovered the man at 635 meters (2,000 feet). He made sure that the tripod under his rifle was sturdy and settled in for the shot. After a breath, his aim rested center on the man's body, and before squeezing the trigger, Bobby asked the Lord to steady his hand. His recoil put him back on target in time to see that he had hit the man. The insurgent's hip was shattered by the bullet. He rolled off the back of the rock and out of sight.

"I just took out a target, three hundred degrees, at six hundred and thirty meters," Bobby reported.

The team commander wanted to be directed onto targets, but Bobby saw more men.

"There are more targets all over the opposite ridgeline," he said.

Bobby was behind the gun, about to scan for more targets, when he noticed that the man he had shot was looking from behind the rock. Bobby aimed for his head, but missed. He fired a few more times to keep the man from returning fire.

In the valley below, the SF soldiers with the ten Afghans were in a fight as well. Bobby scanned the ridgeline again and caught another target; a man with an AK was shooting at the team below. Bobby lasered him at 850 meters. After unloading his AK, the man dropped it and ran to another cache to pick up another weapon. The caches were scattered up and down the ridgeline. Some of the enemy combatants had AKs, while others held RPGs.

In his scope, Bobby had a side view of the man. The enemy climbed up the mountainside. It was so steep that he had to use his hands to crawl up. Bobby's scope could not be adjusted any higher. It was maxed out, so he decided to use a holdover, meaning that he would have to aim high above his target. After estimating the wind speed once more, he held high and to the left into the wind. His crosshairs were above the man's head when he fired, and the bullet landed just high and to the left. The man knew that he was being targeted, and crawled faster. Bobby made a heavier wind adjustment, but his next shot hit low and to the left. Between the two shots, he figured the necessary holdover and wind adjustment and fired once more. The third shot was a killing blow; it hit the man and caused him to tumble two hundred meters (650 feet) down the hill.

Bobby watched the man's lifeless body slide down the hill before he moved on. As he searched again, however, the enemy fighters began to break contact and run into caves. His teammates called for close air support, and they sat on the ridgeline for hours before the mission was finished. In the end, after all the bombing runs by support aircraft and the soldiers' small arms fire, twenty insurgents lay dead. Unbeknownst to the A-teams, the insurgents had been tipped off to their coming. A total of eighty Taliban were in the valley to begin with, but the leaders and others slipped away, leaving a twenty-man force behind to keep the soldiers busy. The team was able to inspect the dead, and the man that Bobby had ranged at 850 meters had been hit under the armpit, sending the bullet through his body and out his rib cage.

For Bobby, the mission marked the end of his third tour, but it wasn't his last. In six years operating in his A-team, he has been to Afghanistan four times, conducting a variety of missions most of which cannot be talked about—on the record that is. Bobby's role as a sniper helped him to gain the experience needed to become a sniper instructor.

FIVE

DUAL DEPLOYMENTS I

THE Global War on Terror presents many unique aspects in the realm of war fighting, the most evident being that it sustains combat simultaneously in more than one country. Afghanistan and Iraq are at the forefront, and some snipers can brag about the differences after having fought in both.

Go Army

Stan was a sniper with the U.S. Army's 101st Airborne Division. On the morning of September 11, 2001, his sniper section from the Second Battalion, 187th Infantry were shooting their rifles on a range. In the pits, Stan pulled targets for the snipers shooting, and just as he'd finished patching a hole, the range control radio went off.

"Hey, guys, go ahead and put the targets in the shack and lock it up. We're going home," said the platoon sergeant.

"Why?" questioned Stan. Shooting was his passion, and he didn't want to miss out on his turn.

"We've been attacked."

As a senior sniper, Stan was serious about training. He thought the radio transmission was a joke and made his way up top to straighten out his teammates. On the trail, however, he noticed Military Police storming the roads and the range, shutting everything down.

Back in their team room the snipers watched a small black-and-white television to learn more about the incident. The video footage caught everyone off guard; the two World Trade Center towers were ablaze, and the image of the planes diving into them was on replay. It took a few hours for all of the information to trickle down, but soon the men knew the details.

Immediately after 9/11 Stan and his platoon, like many from the infantry battalions, were tasked with a security role. The snipers guarded a weapons deposit for some time before reverting to normal training. From then on, the brigade's training was intense, heading into "black cycle." Black cycle meant that if a unit from the 101st were to be called upon for war, Stan's would be the first to answer.

By this time, Stan was a specialist, on his first enlistment.

He had come a long way from high school to the Army infantry, and from being a regular Joe to being a team leader in the sniper section. His first unit was in South Korea, the First Battalion, 506th Infantry, based in the demilitarized zone. Though Stan had wanted to be a sniper since before he joined, it was there in Korea that he learned what it took to become one.

He befriended soldiers from the sniper section, and after a few discussions, they gave him insight about the selection process for sniper school and how to prepare for it. He did not try out for it there, however, because he wanted to better his chances by waiting to return stateside. Getting into sniper school is hard for soldiers overseas; it is a privilege that most units cannot afford to extend.

Over a year later, Stan was at Fort Campbell, Kentucky, as part of the Second Battalion, 187th Infantry, under the Third Brigade of the 101st Airborne Division. He was a Rakkasan, translated in Japanese as *falling down umbrella man*. The name was given to men in the 187th by the Japanese after World War II, to signify the unit's airborne ability. They are capable of deploying anywhere in the world within thirty-six hours, which excited Stan. He was also satisfied to be carrying on a family tradition of serving in the 101st Airborne Division, as many of his relatives had done before him. There, he seized the opportunity to become a sniper when he learned that a selection was coming up.

Physically, Stan was short and stocky. Lifting weights was

a passion he had carried over from high school, and his muscular build helped with the heavy packs. Mentally, he was always focused, always serious, and willing to study when others would not. These attributes helped him to pass selection and to be chosen as a part of the Reconnaissance/Surveillance and Target Acquisition platoon, or R/STA platoon. That first year in the platoon, he faced his first real test in the Army.

Stan's unit and SOTIC, the Special Operations Target Interdiction Course, were located on the same base, and the course had an extra slot, allowing him to walk on. At first, Stan felt overwhelmed; it is not very often that a young infantry soldier is given a chance to attend such a prestigious class. His peers were mostly seasoned Special Forces veterans with badges and patches that Stan could only dream about, but still he was there and he knew that he had to make the best of the opportunity.

Soon young Stan was rubbing elbows with the best of them at SOTIC. He first noticed the professionalism of everyone, particularly the instructors. They were helpful in every aspect and were adamant about guiding Stan toward success, which was different from a regular Army course. Regular Army courses seemed to make everything hard for the students, as if they wanted them to fail, but not the SOTIC instructors.

Instructors humbly explained their real world experiences and sincerely wanted the students to learn the art of sniping. Stan noticed that whenever he needed advice, the instructors were there. They helped him understand the concepts without his having to worry about being disciplined. It made for less stress during his especially favorite portion of training, the marksmanship phase.

His first lesson during marksmanship was "optics magnify errors." This simply means that without the fundamentals of shooting, a scope only proves a shooter's inability. To understand this, the students were restricted to iron sights on their sniper rifles for the first two weeks. With no scope or bipods and only a sling and a shooting glove, the snipers set out to learn the basics. They shot in the standing using just a sling, slow fire, and rapid fire, as well as different shooting positions at each yard line, ending at six hundred yards. In the end, the training paid off, because once they mounted optics, Stan was dead-on.

After graduating SOTIC, Stan was back with his unit, and then 9/11 happened. Within months they learned that the rumor about Afghanistan was true, and his brigade packed and readied for deployment. Stan hoped all the training and all the shooting would pay off, because now he was headed for war.

Trained for War

By the time Marine Corporal Josh Rush deployed to Afghanistan, he had been through extensive sniper training. He'd arrived to the Second Battalion, Third Marines scout/sniper platoon in 2003 and was a PIG (Professionally Instructed Gunman) for two months before earning a seat at Third Marine Division's sniper school on Marine Corps Base Hawaii. The odds were against him, though, as a private first class, and after months of sniper training he went to the school only to fail the first event, land navigation. Luckily the instructors allowed him to finish the course, but in the end, he was not given a certification.

Rush considered the sniper training a bonus to begin with. He had joined the Marines to be in the infantry, and when he found out about snipers, he jumped at the chance. After high school, he enrolled for a semester of college and quickly realized that it wasn't the right thing for him and considered the Marines. Both his grandfathers and an uncle had served in the military, and his uncle had been a Marine infantryman. When Josh told him that he wanted to be a military policeman, his uncle spoke sense into him.

"If you're anything but a grunt, I'll disown you," he said. After that, Rush signed up for the infantry.

After sniper school Rush returned to his battalion. Of course he was punished by his senior snipers for not graduating, but he would make it up a few years later. As soon as he was back, a team from the Marine Corps Special Operations Training Group arrived in Hawaii to instruct in urban sniping. They allowed the entire sniper platoon from 2/3 to attend, regardless if they were school certified or not. It was a rare opportunity that the snipers relished. To add to their basic understanding of sniping, the urban course brought about new concepts of shooting and operating in a city environment. Rush learned shooting from multiple positions, helicopter shooting, timed shooting concerning windows of opportunity, and glass penetration, all in two weeks. By the end of the course, Rush had received news that his battalion was deploying to Afghanistan, and they began pre-deployment training.

For the work-up, Second Battalion, Third Marines headed to California. The battalion stayed in Twenty-nine Palms to familiarize themselves with the desert and to practice large-scale exercises. The sniper platoon, however, traveled north to Bridgeport to attend the mountainous scout sniper course.

The snipers arrived in the Toiyabe National Forest, one hundred miles from Reno, Nevada. The altitude and terrain simulated the environment in Afghanistan, giving the snipers a feel for what they would be up against. There, Rush learned the effects that high altitudes had on his bullet, and how to shoot at extreme angles. Another important factor for operating in small teams was the immediate action drills, how to react if caught by surprise. In the mountains, especially in Afghanistan, where the enemy has home field advantage, being caught in an ambush is likely, and for Rush, this training was the factor that saved his life.

With three sniper courses under his belt, and other

training, Rush was shipped to Afghanistan. His battalion flew into Bagram Air Base near Kabul and met the Marines that they were relieving. The snipers discussed their new area of operations and exchanged stories. The outgoing Marines didn't give Rush an exact picture of what to expect. Some of the snipers had seen a lot of action while others had seen very little, if any. None of that would matter, though, because Rush's experience would be different from theirs, and soon he found a temporary home in the Kunar Province.

A Bump in the Night

In early 2002, Stan flew to Afghanistan. The long flight gave him time to reflect. As the team leader, he wanted his men to make it back safely. His seriousness had been the butt of many jokes, but now that combat was near, he was glad that he had paid attention during training. The brief that they had received before leaving kept playing through his mind.

*We don't know exactly how long you're going to be there—
at least six months. As far as the enemy is concerned, they've
been fighting here their whole lives, so expect a war. They're
good, and don't underestimate them. Don't take anything
for granted, don't embarrass your country, and do what
you have to do to win!*

When the soldiers passed through Germany, they were given ammunition. Stan realized the magnitude of how close to combat he was later in flight, while gazing through the window. Gone were the bright blue ocean and green carpets below. Instead, Stan fell asleep to the sight of desert tan as far as the eye could see.

He was awakened one hour before landing. As the plane touched down, the soldiers were welcomed by U.S. Marines who had occupied the area with rows of tents and a command post building farther off. The airport had been secured by guards and concertina wire, and for the marines, the soldiers were a sight for sore eyes, symbolizing just how close they were to going home.

Within hours, the soldiers were briefed. Their new home was the Kandahar International Airport, just south of the city of Kandahar. The area was the last stronghold for the Taliban, and still a very dangerous place. Al-Qaeda had been driven to the mountains, but they still fought, using guerrilla tactics, hit-and-run style.

Afghanistan was a new world to Stan. The temperature was cold, close to 45°F, with a constant twenty- to thirty-mile–an-hour breeze during the day. At night the temperature dropped dramatically. The environment around them was worse; Stan thought he was in the Dark Ages because the area looked to be the result of a nuclear holocaust. The land was barren, especially near the airport. Hills, natural water beds known as wadis, and flat plains surrounded the airport. Though civilians were only a few hundred meters away, Stan would never come in contact with them. In fact, the only locals he would meet would be men of the Northern Alliance.

Two weeks after arriving, Stan and his partner, Justin, were sent on their first mission. The plan called for a raid on a terrorist safe house in Khost, north of Kandahar a few hours, and Stan's team was to provide over-watch for the soldiers. They were to fly in on four CH-47 Chinook helicopters, hit the house, capture the men, and return home. But from the very beginning the mission was doomed.

On January 28, the raid force departed in the middle of the night. It was cold in the back of the bird, and Stan tried resting during the ride. Night illumination was great and the pilots could see the ground below perfectly. A few minutes passed before the door gunners tested their weapons into a dry lake bed. Stan was up front near the pilots and was crammed next to fuel bladders, which pressed against his knees. He dozed off a few times, but the sleep was temporary.

As the helicopters neared the target area, everyone was on alert. Stan found his gear and weapons, but he was not prepared for what happened next. As the pilots descended, sand, dirt, and debris were picked up by the rotors and engulfed the area. The pilots could not see anything, including the ground. This phenomenon is known as brown-out. With the loss of visibility, the pilots had no clue when to brake and they landed the helicopters at full speed.

Stan was looking to the rear hatch just before impact. He felt a strong jolt and then all went black. The impact had knocked everyone out. Minutes later, Stan opened his eyes and heard screams around him. It was dark inside, but when he turned his head, Stan noticed that a fire was beginning.

"Did we just get hit?" he thought, while his adrenaline pumped.

It was cold, and the right side of Stan's head was numb. His entire body was wet. Then he remembered the fuel bladders that had been pressed up against his knees earlier, and he knew that he was soaked in diesel. As he gained consciousness, he knew that he needed to get out of there.

In his mind, they had been hit by an RPG, and he believed that they were in a fight. His M4 was still attached to his body, as was his chest harness full of ammunition. He did not have time to worry about the other gear.

The helicopter lay on its right side. Stan gripped his weapon and crawled for the rear opening, but he stopped before stepping out. Outside, others had formed a perimeter around the wounded. Stan was on his way out, but with one last glance inside the helicopter, he saw someone else.

He went back and pulled the soldier out by the drag handle on his vest. Stan's head ached and it was hard to breathe, but he still pulled the man sixty yards to the others. Behind him, flames consumed the bird and suddenly a medic approached.

Stan knew right away that he was a Special Forces medic.

His appearance was different from that of the other soldiers, and Stan felt relieved knowing that Eighteen Delta–qualified medics, the best, were out there. The medic grabbed Stan and shined a light on his face.

"You need to get on that helicopter over there," he said.

The last thing Stan wanted to do was get back on another helicopter. His face was swollen and bleeding, and he could not open his right eye because the side of his face had been crushed. A piece of metal from inside the helicopter had slammed down on his face and into his eye socket, tearing muscular tissue. It was fortunate the metal stopped when it did, because less than an inch farther, it would have gouged his eye out.

Stan obliged the medic and was evacuated. Shortly afterward, he was flown to Germany for reconstructive surgery. Weeks later, the surgeon asked if he wanted to go home or back to combat. He could not believe that his first mission had gone the way it did, and he hoped that it would not be a sign of what was to come, because he had decided to get back into the fight.

Kunar

In 2005, the Second Battalion, Third Marines unit entered the Kunar Province. This northeastern region of Afghanistan would prove to be beautiful and untamed. Bordering Pakistan, the province let fighters move to and from the adjoining country, and the mountains in the region made it hard for U.S. forces to stop them. It was June, ushering in good weather and more enemy activity from Anti-Coalition Militias (ACMs) who emerged from their winter caves now more willing to fight. ACMs were bands of fighters with ties to the Taliban and al-Qaeda, but not specifically part of the groups. They were typically tribes or groups who were not fond of the Coalition and had lived their entire lives in Afghanistan, giving them an edge over foreigners.

Rush and his team joined Echo Company at Camp Blessing on the western front. He was clueless about the country, but it didn't take long to adapt. The small, remote base sat at the bottom of a valley surrounded by mountains. Atop the mountains were observation posts held by Afghan Security Forces. The base itself contained a few buildings surrounded by concertina wire and dirt barriers.

For operating, Rush trusted that his team was prepared. Sergeant Evers, the team leader, was very experienced and knowledgeable. He'd been through sniper school in the early 1990s and had several deployments, including a trip to Iraq. He was also older, in his mid-thirties, and had been an instructor at the mountain warfare center in California, teaching snipers to shoot, navigate, and survive in the mountains. If anyone knew how to manage in the peaks of Afghanistan, it was Sergeant Evers. Rush's other teammate was Parish, a corpsman who had recently joined the platoon. He was quiet, but hardworking and listened well.

At Camp Blessing, the snipers were given time to adjust to their surroundings. They climbed the steep hills and stayed at the observation posts a couple of times to feel the effects of the elevation. Sergeant Evers was in great shape; he ran faster than anyone in the platoon, and when it came to humping, the more weight he carried the faster he was, but Rush was behind him all the way, in great shape himself. Parish, though, being fresh to the platoon, hadn't experienced anything like Afghanistan, and he sat the first few missions out. During the adjustment missions, Evers was behind the scenes requesting for the snipers to run solo missions, and one night they were approved.

From the week they arrived, the marines took enemy fire. Their camp was hit at least twice a week by mortars and small arms. Evers wanted to find the enemy's attack positions, but the battalion was hesitant to use the snipers. Finally one night the camp commander, a first lieutenant, knocked on their door and informed them that their mission was approved.

Their first mission was a good learning experience. Evers planned to recon "Rocket Ridge," the hillside from where the marines suspected the anti-Coalition militants shot from. The snipers packed for a three-day operation, with food, equipment, and twelve quarts of water, and set out. The climb to the top of the nearest hill was exhausting and the heavy packs wore the snipers out, but they were able to make it into position. When the sun rose, Rush noticed the craters from return attacks by artillery. The dense forest hid small trails, and the most interesting thing the snipers found were early warning devices set by the fighters. The enemy had placed piles of dead branches on the trails, and if the snipers had moved through them, the sound would have been loud enough to alert the fighters.

After the first few missions, the snipers gained confidence. They took to the mountains and began to understand the region more. The only contact that they had made with the enemy was minimal; it happened when their base was attacked. The enemy fighters would not stay long enough for the marines to pursue them, or for the snipers to maneuver on them.

Meanwhile, battalion headquarters began planning Operation Red Wings. The mission was to kill or capture anti-Coalition militias in the Korengal Valley, a place that seemed untouchable to U.S. troops. The Marines wanted to clear the valley of the fighters before the upcoming Afghanistan elections. The main anti-Coalition militia, known as the Mountain Tigers, was led by a local Afghan Arab known as Ahmad Shah. He kept a tight grip on the Korengal, and became a key target. Finding him and his fighters, however, would be a task in itself.

Disaster met the start of Operation Red Wings. A four-man sniper team made up of U.S. Navy SEALs, which coincidently would have been Rush's sniper team, if the marines were given the mission, infiltrated deep into the valley, but were compromised and subsequently attacked by Shah's men, who killed all but one, Marcus Luttrell, who escaped. The rescue team sent to the aid of the men were hit while in their helicopter, and all were killed. As a result, a month later the Marines launched Operation Whalers.

Anaconda

In Afghanistan the action was heating up. Back at Kandahar, Stan was relieved as a team leader because of his absence and his injuries. That was unimportant to him; what was important was that he would be fighting soon. Fortunately three days after he arrived back, he learned that they were going to face the enemy soon, as a big operation was at hand. The soldiers were told to prepare, but the specifics were vague.

In the meantime the snipers fine-tuned their weapons. In that climate, elevation and weather have effects on the bullet, and the snipers knew that. They trained to anticipate their weapons' efficiency by shooting as often as they could, which was every day. Stan became the spotter for his new teammate, Jason. They were able to quickly mesh as a team, and Stan took to his new role well. Within a few days, the snipers were ordered to pack because they were headed north.

As for Stan, he faced his fear for the first time since the wreck by getting on another helicopter. The flight seemed like forever, but soon the soldiers arrived twenty-seven miles north of Afghanistan's capital city, Kabul, to their new home, Bagram Air Base in the Parvan Province. The base was an old Russian airstrip used by the Soviets in Afghanistan, but now it was property of the U.S. military.

Shortly after settling in, the snipers were told of the next big mission; it was dubbed Operation Anaconda. They learned that al-Qaeda fighters were gathering in the nearby Shah-i-Kot Valley and had been doing so for months. Intelligence reports stated that a few hundred fighters, including high-value targets, were in three villages along the valley. The fighters were also preparing a spring offensive against Afghan and U.S. forces, but the plan was to hit them before they could attack.

For the soldiers, the overall plan was fairly simple. Task Force Hammer were Afghan military forces accompanied by U.S. Special Forces troops responsible for pushing through the valley, essentially destroying the enemy or driving him toward the second element of the plan. Task Force Anvil, the second element, were soldiers from the 101st and Tenth Mountain, with support from Canadian troops, and they were to wait in blocking positions on the northern and eastern edges of the valley to prevent escape and deny enemy reinforcements.

Stan's team was to insert with the ground troops on the eastern ridges. Once there, they were to break off and move around a mountain pass to cover the infantrymen's blind spot. The soldiers were told that it was to be a two-day mission and to expect some contact but not much.

Before the assault, Stan and Jason learned how dangerous the mission was. They were zeroing their rifle one last time when they ran into a Special Forces soldier with whom they had gone to SOTIC.

"Why are you carrying such a big weapon?" Stan joked, referring to the light machine gun slung around his old friend's neck. "Doesn't that slow you down?"

"Well, where you're going, the only thing they run from is a belt-fed weapon," said the SF soldier, knowing what Stan and his partner were up against.

He told the two snipers about the terrain and how rugged it was in the valley. He mentioned that the elevation alone was over nine thousand feet, and as if these two factors were not enough, the SF soldier went on.

"Remember, they have the home field advantage. They have every possible route mapped out. They've been living in that area for hundreds of years and have fought off invaders dating back to the times of Alexander the Great. It's going to be a hornet's nest in there, so just be ready," said the soldier.

Although he was grateful for the information, the talk didn't inspire Stan to be brave. In fact that night he did not sleep a wink. The very next morning his team and the other soldiers moved into the Shah-i-Kot valley.

Technically, Operation Anaconda began on March 1, 2002. Special Operations teams infiltrated the valley to begin surveillance on and reconnaissance of landing zones and enemy target sights. On the morning of March 2, as dawn approached, the first wave of birds loaded with soldiers departed Bagram Air Base. Aboard, Stan's team rode in with a rifle company and were not quite sure what kind of fight they would be facing. Whatever happened, Stan trusted Jason, his friend and partner. The two of them had been in the platoon together for a long time and had attended the same class at SOTIC, which added to their bond. Stan was also glad that Jason was a great shot.

AH-64s, the U.S. Army's premier attack helicopter, led the way into the valley. They covered the more vulnerable CH-47s loaded with soldiers. The helicopters entered the Shah-i-Kot valley and hoped for the element of surprise, but instead, the door gunners immediately went to work because they were being fired upon.

The sound of the helicopters alerted the enemy from caves and bunkers in the mountainside. They unloaded with machine guns and AKs, hoping to down the choppers. Stan and Jason sat anxiously near the rear ramp, close to the manic tail gunner who was giving the enemy hell with his M60 machine gun.

They flew in low, and Stan saw people gazing at them from the ground. When he turned his head, the bird jerked, and a stream of white smoke rushed past the back ramp. It was a rocket-propelled grenade. The soldiers were given the warning for one minute out, and the two snipers prepared to land. Suddenly, Stan felt one of the crew chiefs grab his shoulder.

"Primary landing zone is hot! We're moving to the alternate LZ, pass it on!" he yelled.

In his mind, Stan remembered that the alternate LZ was only a stone's throw away, meaning that it, too, would be hot.

Around him, everyone prepared to exit. It was a mad dash to gather gear and get packs on. Stan and his partner stood up just in time to see the machine gun team opposite them shuffle toward the exit. When they stepped off the back ramp, it was evident that the bird was hovering higher than expected. They were around ten feet off the ground. When Stan stepped off, he fell hard into the dirt below but was not injured.

With the last man off, the helicopters left, along with their loud clatter. The soldiers were at the base of a mountain range, preparing to move. Around them, firefights ensued from the other LZs. As the spotter, Stan was on the radio and put the handset to his ear to hear the traffic. Jason prepared his rifle, while Stan listened to the brigade network and heard that others received contact upon insert. He switched to the battalion network and learned that all of the elements from the Tenth Mountain were in serious fighting.

Suddenly, shots rang out. Everyone hit the deck and scanned the area. Stan saw something move in the distance, around two hundred yards away. At first he could not make out what it was because it disappeared behind objects. Finally he got a good picture of a man with the signature brown Pakol hat dubbed "pizza hats" by the soldiers. The hat was indicative of local fighters, and gave away the shooter's position to Stan.

Stan crawled to a boulder close by to get a better view. His partner was handling the rifle, and the man was closer now, around 150 meters, in clear sight. The fighter rested on one knee and took aim at the soldiers near Stan. Without warning, Stan sighted in with the M68 Aimpoint scope mounted on his M4. Stan rested in a modified prone position and set his sights center on the man's chest. But after two quick shots, the man still did not go down.

"Did I miss?" he questioned.

"What do you have?" asked Jason, but Stan did not have time to explain. He glanced back toward the enemy, who was now rushing for him. Immediately, Stan aimed and squeezed off another shot, and hit the man high in his chest, which put him down.

Seconds later the soldiers began to move. As they climbed, Stan figured that he had shot what was to be the first of many fighters, but he also learned that what the commanders had said was true. The altitude was also the enemy.

Their objective was 1,000 kilometers (33,000 feet) up the mountainside. The climb was taxing, especially with ninety pounds of gear. At that altitude it was hard to breathe. Stan and his partner were flanking the platoon to prevent ambushes. They used a few small goat trails that allowed for an easy trudge, but unfortunately the trails did not last. Most of the climbing was with hands, moving nearly vertical.

After some time the men finally reached the top.

"Did you shoot that guy back there?" asked Jason.

"Yeah," replied Stan, still somewhat out of breath.

Gunfights echoed throughout the valley as the soldiers took positions. Although they'd passed nine thousand feet in altitude, there was still plenty of mountain. The snipers found a position to shield them from above and from across the valley.

Stan and Jason decided to stay with the rest of the soldiers and dig in. Soon the enemy fighters were onto Stan's element. Other soldiers began taking fire from across the valley, and a machine gun team near Stan's position received contact.

The two snipers were dispatched to help them, and they scrambled one hundred yards below them to the machine gun team.

"Where's the fire coming from?" asked Jason.

The soldiers pointed the snipers in the right direction, and Stan used his spotting scope to comb the area. Snow covered some of the rocks and pine trees in the distance, and there were a lot of shadows coming off the landscape. Stan was not able to find the shooter, but fifteen minutes later they all heard the familiar crack of a bullet passing overhead.

Stan immediately examined across the valley again. After

another empty search, he narrowed the enemy's location to three possible areas. From the angle of the passing bullet, Stan knew that the shooter was above them, possibly near one of the tree lines. There he set his sights and scanned back and forth.

The shooter was silent for close to forty-five minutes before he made his last mistake. He took another shot and Stan spotted his muzzle flash. The man was lying in a hole surrounded by snow, with only his weapon and the top half of his body showing.

Just moments before this, with help from the machine gunner, the snipers had been able to collect somewhat of a wind call. Stan asked the soldier to let out a burst of fire from which Stan would spot the tracer round. The soldier obliged, and Stan was worried after seeing the wind take the bullets for a joy ride. Their only chance of a hit would be to shoot between wind gusts.

After the enemy fired, Stan quickly directed Jason on target. The enemy was around 750 meters, or a half a mile, away, and Jason dialed his scope and waited for the wind to die out. As soon as it did, Stan spoke up.

"You need to shoot immed—" he said, but Jason was already on the ball and let loose before Stan could finish his sentence.

The shot was timed perfectly, and Stan watched the bullet impact only eight to ten inches to the right.

"Hold left, just off his right shoulder!" he said.

Jason did and fired again. The second shot was right on target, hitting the man low, near his stomach. The man slouched forward and stopped moving.

"Great shot!" said Stan.

Two minutes later, though, the man was shooting again. Stan looked at him, surprised that he was not dead. Jason had inflicted a wound, but it did not kill the shooter instantly, though he would eventually bleed out.

"Put another one in him," Stan announced.

Jason shot again, but this time the round landed short, just in front of the man's position. The bullet had looked to be dead-on and Stan guided Jason to shoot again. Sure enough, the last bullet put the man down for good.

Afterward, Jason and Stan moved back up the mountain. Across the valley, the plan took an unexpected turn. Task Force Hammer, the preselected main element, had been compromised in the early stages of the operation. They were no longer able to sweep through the valley, and now Stan's element, Task Force Anvil, was the main force.

Later that afternoon, Stan's team and the other soldiers were instructed to break position and fall back down the mountain. The battalion wanted everyone to reconsolidate and move to different locations. The soldiers waited until night to move. It took four hours to navigate down the dangerous terrain. Stan hoped not to trip, because there was no telling how far he would roll. Once down, the soldiers took up a position in an empty wadi for the night.

Stan could not believe the cold. It was near –20°F with the windchill. The snipers had one sleeping bag between the both of them because they expected a two-day mission. Throughout the night the two traded the use of the sleeping bag, but the cold, the rocks, and sporadic mortar fire made it hard to sleep.

Before sunrise, the soldiers prepared to move. They were to move to position Amy, one of the pre-established blocking positions. It was a few kilometers away and they needed to move fast. Around 10 A.M., as they were hugging a bank near a wadi, Stan saw one of the soldiers break rank and walk toward the small stream of water a few yards away.

"Don't go over there. You're probably going to get shot," someone said.

The soldier did not listen, and as soon as he bent down to wash his face, they were attacked. The enemy opened up with a barrage from a 12.7mm Dushka machine gun. It, along with mortars, began to smother the soldiers.

Everyone hugged the embankment except the soldier near the stream. Bullets hit near his feet and he did not know where to run. Though it was not funny, Stan could not help but laugh at the soldier. He looked like he was dancing. Miraculously, the soldier made it to the side of the wadi and out of danger. When he did, the shooting eventually stopped, allowing the patrol to set out again.

A few hundred yards later, the wadi ended and the soldiers climbed out. Suddenly a pop rang out in the distance. Stan glanced to the front of the patrol and saw an RPG land fifteen feet away from him, at the feet of another soldier. The blast threw the soldier onto his back and also knocked Stan and Jason over. Stan was dizzy for a second but recovered, and noticed that the soldier who had been hit was severely wounded. Shrapnel had mangled his legs and ripped into his left ribs and armpit. The impact had destroyed his front Small Arms Protective Insert (SAPI) plate and all of the magazines within his chest rig.

Instinctively, Stan and Jason ran over and dragged the unconscious soldier into a nearby ditch. Another volley of mortars fell around them, while others found their way into the ditch as well.

"This sucks! Every time we turn around, we're getting attacked!" Stan thought.

The hole they were in was fifteen feet deep. Above them, mortars pounded the banks, sending dirt in every direction. A medic tended to the wounded soldier, but the damage was bad.

Everyone waited for the barrage to stop, but it did not. Mortars landed close, shaking the ground in the hole. Stan knew that if one landed in the ditch, they would all be dead.

Suddenly the inevitable happened, one landed inside. It hit high. Amazingly, nobody was injured, but dirt flew everywhere.

The two snipers looked at each other. Stan realized there was the possibility of a hand-to-hand combat with the enemy.

"What do you think?" he asked, wondering if they were going to live.

"Man, I don't know," replied Jason.

"This is what is going to happen. We will continue to watch each other's back, and either we will end up crawling out of here tonight, or they're going to send a group in to close with us. Maybe we'll end up fighting it out right here. This could be the end," Stan said.

They were exhausted from the movement and from the sleepless night. Stan hoped to have the energy to fight with his hands if he had to.

Then Stan made a foxhole promise.

"If God can get me out of this, I'm not going to be such a jerk when I get back," he thought. In the States, Stan was a loner. When his teammates opted for a night on the town, Stan chose to hit the gym. When they wanted to catch a movie or hang out, he declined in favor of going to a local shooting range and practicing his technique. But in that hole, he promised God that if he made it out, he would be a changed man.

The soldiers were trapped in the ditch for hours. A few minutes after sundown, however, the mortars let up. The enemy mortar team had been on a hill known as the Whale because of its resemblance to a whale. By nightfall, a recon team had spotted the enemy team and called in air support, killing the fighters.

That night, the soldiers made it to their position and settled in. When Stan was awakened for watch, the valley was cloudy, dark, and very cold. Sporadic fighting could be heard throughout. With his night vision, Stan noticed two MH-47 helicopters, used by Special Operations, flying overhead. He wondered what their mission was as they flew deeper into the valley.

"It'd be nice if a bird came and got us. This sucks!" he thought.

Ten minutes later, through his optics, night turned into day from explosions in the distance. He grabbed the radio and turned to the brigade network, but did not hear anything immediately. Then he heard that one of the helicopters had been shot down, and that the fighting was intense. Stan remembered what his Special Forces friend had said back at the Bagram Air Base, and it was the truth. It was a hornet's nest.

The next morning, Stan's unit was tasked to move a short distance away. A mountain between them and their destination, along with the brutal terrain, caused the movement to last from 0900 to 2200 that night. Along the way, the soldiers ran into a Special Forces team on all-terrain vehicles coming down from where the soldiers were going. When the Special

Forces team learned of their intentions, some of the men chuckled. Stan overheard them say that another 12.7mm Dushka was wreaking havoc in the area and that the soldiers shouldn't try moving down the other side of the hill. They also gave them grid coordinates to a cache of extra MREs and five-gallon water jugs that they had left behind.

When they reached their position, Stan and Jason teamed up with four recon soldiers from the platoon. Their job was to observe for enemy reinforcements moving into the valley and to destroy them. The six of them broke away from the other soldiers and climbed for most of the night. The trek was literally breathtaking, forcing them to stop every few hundred feet to catch their breath. When they were high enough, the team positioned themselves where the Special Forces soldiers had been.

After daybreak, the men sat somewhere between thirteen and fourteen thousand feet altitude. Stan realized that he was looking into Pakistan. From then on, the team called in air strikes on vehicles carrying reinforcements and on enemy mortar teams. Insurgents kept on the move, and the team used optics to spot them. When they did, their weapon of choice became the deadly A-10 Thunderbolts, known as Warthogs. This aircraft specializing in close-air support brought with them bombs and 30mm Gatling guns firing depleted uranium armor-piecing shells at a rate of 2,100 to 4,100 rounds per minute.

Meanwhile the other sniper teams were seeing action of their own. East of the Shah-i-Kot Valley, in the town of Khost, snipers from Stan's unit were in heavy fighting, too. Many of the al-Qaeda fighters that Stan and his men encountered had moved through Khost, fighting their way past the U.S. soldiers there. On one occasion, the insurgents almost made it into the soldiers' compound. The spotter for the two-man sniper team, a soldier named Vinny, shot eighteen high-explosive 203mm grenades at one mud hut. Later when the hut was inspected, the soldiers found body parts, a scalp, and a sandal. Jason, the shooter, went through most of his ammo for the bolt-action M24 in a matter of hours. In the length of time they were in Afghanistan, those two snipers accumulated more kills between the two of them than most of the entire battalion.

Elsewhere in the valley, other snipers made history. Attached to the 101st were Canadian snipers from the Third Battalion, Princess Patricia's Canadian Light Infantry. Two three-man Canadian sniper teams infiltrated the valley along with U.S. soldiers. For days they provided covering fire for U.S. soldiers and killed targets at amazing distances. One soldier, Corporal Robert Furlong, armed with a .50-caliber sniper rifle, dispatched an enemy fighter from a distance measured at 2,430 meters (8,000 feet). It was a world record shot.

After nine days of calling in fire support, Stan and his team climbed off the mountain. Two uneventful days passed at the bottom of the hill before they were flown to Bagram Air Base. The simple two-day mission of Operation Anaconda had turned into twelve days, and in that time Stan went from 185 pounds to 158 pounds.

At the close of the summer, Stan and his partner were flown back to Kandahar. Their stay in Afghanistan was coming to an end and the Eighty-second Airborne was coming to replace them. Stan was glad that he had been to combat, but he was unsatisfied that he had not been able to actually get behind the sniper rifle and do his job. In the final weeks of his stay, he learned, however, that he just might get the chance to do so. The rumor was that a war with Iraq was coming down the line, and all of the 101st Airborne Division was going to be a part of it.

Korengal Valley

Operation Whalers called for the marines to enter the Korengal Valley and clear it of Shah's men. Among the battalion, two companies would hold blocking positions to intercept fleeing fighters, while the rest of the marines patrolled into the Korengal. Rush's sniper team was picked to lead the men patrolling as over-watch for the second group.

To start the mission, the marines assembled at the mouth of the valley. Mortar men pre-set target points while infantrymen prepared their weapons. Having already packed, Evers, Rush, and Parish arrived at night and immediately moved out. When Rush had entered the Korengal before, they had been hit. He knew that ACMs ran wild in the region, and he expected a fight.

The three snipers departed under a serious mood. They wanted payback for the SEALs and were angry at having to do another seemingly endless trek through the mountains.

On the first night, Rush and his team made it to the top of the nearest mountain and rested at the edge of a large grass field. It took them most of the night to reach the peak. They were fully camouflaged in green blouse tops to match the vegetation of trees and brush above the ground, and desert trousers and boots to blend with the dirt and rocks. They had also painted their faces.

Though they were loaded down, they carried only standard equipment. Rush had an M16 with sixteen magazines, a range finder, spotting scope, tripod, binoculars. Evers had the M40A3 and an M4 with two radios. Parish had an M16, his medical kit, and explosives. By now, they were conditioned to carrying such heavy packs, and patrolling with them was easy.

They waited the rest of the night, and in the morning the push began. For five days, the snipers acted as the eyes for the marines on the valley floor. Before the marines moved, Rush and Evers used their binoculars to scan the hillsides and lowlands for fighters preparing to ambush the team. Rush remembered the observation skills taught in sniper school; the school had prepared him for what they were now doing.

In that time the snipers took on a hunter's mentality. They realized that it was the enemy's home turf, yet they weren't crippled with fear. Moving through the trees and along rocky trails, they observed target areas for the battalion, and by the end of the first week, they were deep into the heart of the valley, at the base of Sawtalo Sar.

Shah's men lived on the valley's highest peak, Sawtalo Sar. Rush's sniper team arrived at the base of it before the company and waited for the infantry to meet them. After a quick rendezvous, surveillance aircrafts reported that eighty of Shah's fighters were trying to escape. When he heard that, Evers told the company commander that his snipers would probe the village where Shah's men were holed up. It was a plan, and the snipers resupplied themselves with food and water and then began to climb the mountain.

It was a difficult march for the snipers. Rush was slightly paranoid knowing that if they made contact, it would be eighty versus three. The region was Shah's safe haven and the place where the Navy SEALs had died. They climbed for most of the night, but stopped to rest. As a precaution, they emplaced claymores around them while they slept. Rush did not want to sleep, so he took three No-Doz pills to stay awake and let his teammates rest all night. Early the next morning, they moved again, and made it to the mountain only ten minutes earlier than the company who had been moving all through the night.

When the marines arrived, they stormed the hilltop without resistance. The company set out a perimeter allowing others to investigate the structures and debris left behind by Shah's men. Rush noticed the wreckage area of the helicopter carrying the SEALs and the fighting positions that Shah's men had used. Foxholes, barriers, and hides covered the mountainside.

All the while, Evers talked with the company commander about their next objective. The marines were moving back down the hill and out of the immediate area, requesting that the snipers cover their movement.

"Hey, guys, they're going to take off soon," Evers informed his team. "We're going to stay on the ridgeline and provide over-watch for them," he continued.

Rush knew the plan. The snipers would continue to be advance warning, and shortly after, they threw on their packs and moved out. Along the way, Rush observed the area for the marines, and after they moved past a clearing, the snipers unexpectedly ran into a single man. He appeared to be a goatherd, but with no goats. Evers searched him, but he was clean of any communications devices. Parish took his picture for identification and they let him go in the opposite direction.

"I just got a really bad feeling from that guy," said Parish, "I could see it in his eyes."

Rush thought nothing of it as they moved forward. As a safeguard, Evers told Rush to use the spotting scope to make sure they weren't going to walk into an ambush. The threat of eighty fighters still remained. A thorough search revealed nothing and the snipers pressed on. They patrolled as they always had; Evers liked to walk point, with Parish in center, while Rush took tail end Charlie.

Evers stressed stealth. The snipers always moved slowly and precisely, careful of the noise they made. It was a fundamental element for the success of such a small team. Five minutes after releasing the goatherd, the three snipers crept down a small hilltop before entering an open field with waist-high grass. There, suddenly, gunfire opened up on them.

From the other side of the field, militants shot at the snipers. Rush knew nothing but to react. Bullets cracked near his head and arm, narrowly missing his body. He raised his rifle toward the militants and threw a long burst at them. In a flash, he remembered the immediate action drills that he had done for physical training every Thursday back in the States, and also during the mountain sniper course. His reactions were etched into his brain.

In front of Rush, Parish unleashed his M16. However, the first volley of AK fire from the militants sent one bullet into Parish's leg and another into his chest, knocking him onto his back. His night vision goggles and magazines stopped the second bullet, but his leg was useless. The bullet had entered below his knee and ricocheted up, tearing muscle tissue and eventually breaking his femur.

The meeting was chance contact between the two groups. Nearly twenty militants were not expecting to find the snipers at such close proximity. In passing, they noticed the snipers and were the first to fire.

After firing, Rush turned and bounded back as he was taught. With his back to the firefight, he felt two bullets hit his pack. Luckily they didn't hurt him, but he knew what he needed to do next.

"Drop your pack so you can move faster or you're going to die," he thought to himself.

Steps later, he fell on his knee, pulling the quick release straps and flinging his heavy pack to the ground. At the same time, he thought about his teammates. He heard them shooting, but the enemy fire outweighed theirs. Instinctively he turned and returned fire.

Rush dumped an entire magazine before turning to run. He was on his feet for a couple of seconds before two bullets slammed into his back SAPI plate. It felt like two sledgehammers smashing into his back. The impact shoved him forward, but he didn't fall; instead he took a few more steps before spinning to return fire once more.

The militants didn't let up. They continued to fire, but the snipers answered well. Rush and Parish were not aware that immediately after contact, Evers had dropped to the ground, moved to cover, and called in mortars and artillery, even after it was deemed danger close.

Rush's adrenaline kept him shooting and moving quickly. While running back once more, he was shot again in the back, but his SAPI plate held strong, keeping the third bullet from entering his body. After the impact, Rush anticipated his own death. After not hearing or seeing him for a while, he assumed that Evers had been killed, and he knew that Parish was injured.

When he stood up, immense fire rained in on him. Fifty meters away, the militants yelled at one another. Rush clearly heard their voices and remembered the video of them killing the Navy SEALs, but he was determined to fend them off. A quick check showed that he still had plenty of ammunition, and with that, Rush moved back again. He was just about behind the small hill that they had walked down. There he decided would be his last stand.

At the same time, Parish had crawled around the hilltop as well. Rush was resting on his knee, looking for the militants, when he smelled a familiar scent. Back home in Michigan's Upper Peninsula, he had hunted deer as a kid and he remembered the stench of gushing blood while gutting them. That was the smell, except it was not a deer he smelled, but blood from Parish's leg.

Parish was severely wounded and quickly losing massive amounts of blood. Rush had no choice but to set a tourniquet on his leg. The damage made Parish pale and left him in excruciating pain; Rush could see it on his face. Next, Rush thought about Evers and called out his name.

"Evers!" he yelled, but to no avail.

Evers had finished calling for fire and was crawling to Rush's voice. Rush was just about to go look for Evers, when he saw him coming. It was a complete relief.

"Keep your heads down, we've got support and it should be dropping soon," said Evers.

Evers, a qualified emergency medical technician, quickly looked over Parish. He and Rush dragged him to a more defendable position, and Parish did not look well.

"Do you want morphine?" asked Evers, but Parish declined. He wanted to be conscious when the men tried to overrun them.

Rush knew that he'd been shot, but it had not dawned on him that the SAPIs had saved his life. He felt as though he might be in shock, but when Evers inspected him, only small amounts of blood were lifted from his back.

Suddenly, their lifelines were extended. Artillery and mortars began to fall, but they were close. The snipers lay down while explosions rocked the immediate area. Rush felt balls of dirt and rocks hitting him. At the nearest break, Evers remembered his pack had the sniper rifle strapped to it. He had dropped it before crawling away. He was not going to let the militants make off with his gear and he made a dash to get it.

All the while, the other marines rushed to help the snipers. Once they heard the shooting, a platoon commander grabbed ten marines and ran to the snipers' position, arriving just after the fire support had finished.

Evers retrieved his pack, only to turn around for Parish's, which contained the medical equipment. With it, he helped Parish while Rush kept security. One of the other marines called in a medivac.

When it was all over, the marines swept through the area and were astounded. Not a single piece of brass was left from the enemy's position. They had cleaned up everything except two small shards of a rag with blood on them.

Soon, a secure landing zone was formed. Parish took morphine and was attached to an IV. Two Apaches held security for two Blackhawks to pick up Parish. Rush was ordered to go as well. He was excited to get a ride from the valley, and that he'd soon be sleeping in a bed after eating hot chow. He did feel bad for the two poor souls who were forced to carry his and Parish's packs after they were not allowed on the helicopter.

In the end, Parish was medically discharged. Rush was cleared of medical injuries soon and went back to sniper operations. After the incident, restrictions were set on the snipers. They weren't allowed outside the wire with less than six men, and they needed a squad security element within ten minutes' distance.

For Rush, his sniper operations in Afghanistan were finished. Soon he would be back home and on his way to another war, a few thousand miles away in Iraq. He had gained respect for the enemy fighters in Afghanistan. They fought hard and with skill. On his next tour, the enemy he faced would not be much like their counterparts in Afghanistan. They would be much easier prey.

SIX

AREA OF OPERATIONS: IRAQ

WITHIN two months of arriving to Iraq's western Al-Anbar Province, Sergeant Santos, a Marine sniper, learned the dangers of traveling there. By October of 2004, roadside bombs made very few things more dangerous than navigating the country's roads. The IEDs used by insurgents put all convoys in jeopardy, but one night Santos eluded certain death by twenty feet, and a week later he returned the favor to the insurgents.

It started when he and his partner Steve joined two more snipers for an observation mission. Their task was to find and eliminate those responsible for planting bombs near a certain intersection. Insurgents favored that particular crossing and were able to plant IEDs there for months with no repercussions.

For insertion, the snipers rode with MAP Three Squad, from the battalion's Mobile Assault Platoon. To avoid mass casualties, they dispersed into separate vehicles. To determine who won the most dangerous ride, a seat in the lead vehicle, the snipers normally played rock, paper, scissors, but this time Santos took it instead.

Experience had taught the MAP squad to travel with infrared lights and night vision goggles to avoid detection by insurgents. After a quick forty-five-minute drive, they arrived at the intersection. Santos's Humvee stopped twenty feet from the crossing, and the marines dismounted to search for IEDs.

Potholes and craters from previous explosions lay scattered about. Santos stuck close to the squad leader, and they both checked different holes, knowing that insurgents used them to hide mines or IEDs. When they reached the largest hole, in the middle of the road, the squad leader shoved a pencil into the dirt at an angle and hit something.

"Check this out," he said to Santos, who took a knee beside him.

Wiping away the top layer of dirt, the marines uncovered two green antitank mines stacked on each other. Had they not stopped, their vehicle would have been the first to get hit, and with that amount of explosive, they would have surely died.

It was a wakeup call for Santos. Right then, he made it his personal mission to stop the roadside bombs there. That night, as MAP Three waited for explosive ordnance disposal technicians to blow the mines, Santos, his partner, and the other team members hiked to a hilltop nearby to keep eyes on the area.

The next morning, they realized that the elevated position gave them a commanding view. From left to right, the open desert stretched as far as their eyes could see. A civilian road spanned east to west, eventually running along the base of their hill only 100 meters (330 feet) below them. Another road, strictly limited for Coalition Forces use only, ran north to south, and where the two roads met marked the intersection at eight hundred meters (half a mile) from the team.

Two days later, around noon, Santos's partner, Steve, was on watch. Santos and the other team leader, Adam, rested while the last marine, Anthony, held radio watch and rear security. Shortly after he took over, Steve watched a vehicle pull into view.

"Hey, a truck just stopped," he informed his team. Normally, civilians slowed at the intersection to avoid colliding with military vehicles, but stopping was strictly prohibited.

When the snipers heard this, they all reached for optics. A large semitruck was parked on the east-west road, and the driver quickly jumped from the cab and sprinted to the pothole where the mines had been found. There, he dug into the dirt, and once he realized that the mines weren't there, he ran back into his truck.

As team leaders, Santos and Adam were forced to make a decision. The man obviously knew about the mines, but he didn't appear to have weapons. They decided that it was best to stop the truck anyway.

"Engage him," they said.

That's exactly what Steven wanted to hear. He quickly pulled the .50-caliber M-107 in front of him and aimed in. They had previously ranged the intersection and preset their scopes to fit the distance. Two months out of sniper school, Steven would now be putting the training to use, but by the time he had settled in for a shot, the truck was driving away.

"Aim below the engine block," said Adam, but the truck's velocity caused Steven's first shot to hit the trailer attached to the back of the semi.

At that speed, sniper rifles were useless.

"Cease fire," said Adam, patting Steven on the back. They figured that the driver wouldn't have been able to see the one shot, and hopefully he hadn't realized that he was being targeted.

After reporting the incident, the team stayed in place. Steven beat himself up over the shot, but Santos explained that almost anyone would have missed. Around dusk, the snipers prepared to extract. In a few hours a MAP squad would take them back to base, and while they packed, Steven took watch again and noticed that a car had stopped at the intersection.

"Guys, we have another vehicle," he relayed to the others.

This time Steven had the SASR handy. When the snipers looked toward the vehicle, they saw two men jump from the backseat and run to the side of the road. They carried shovels and dug, while a third man carried two objects toward them; they were artillery rounds fashioned into IEDs.

"This is it," said Santos. They didn't have time to report the action; they had to engage before the men escaped. He asked to use Adam's M40A3 lying nearby.

"Go ahead, I'll spot," said Adam.

Steven planned to disable the car while Santos targeted the men. The two of them lay side by side and waited for Adam to count down to fire simultaneously and minimize the insurgents' ability to locate them.

"Five, four, three, two, one" counted Adam, and the snipers opened fire.

Santos, however, laying next to Steven, hadn't anticipated the back blast of the .50-cal. It threw dirt into his eyes and knocked his rifle over before he could fire. When he recovered, he quickly picked his rifle up and scooted to the left, out of the Steven's dust.

When he found the car, Steven's bullets sent sparks into the air like fireworks. The driver realized the situation and started the vehicle. Santos swung his rifle to the right, searching for the men digging; one stood completely still, in shock. He was Santos's first target.

Wiping dirt from his face, Santos concentrated. His crosshairs bounced momentarily until he slowed his breathing and relaxed. The man in his sights turned his head toward the car, just as Santos let one fly, but his shot missed. It was close enough to make the man duck, but instead of running for the car, he ran in the opposite direction.

Steven shot the car three times before it started to move. All of the men except the one who had run were in it. They had no clue where the snipers were and drove straight for the marines' hill.

With a quick search, Santos couldn't find the man who ran off. The lucky bastard had made it into defilade, and Santos turned his attention to the car.

"Get on line," said Adam, realizing that the car was going to drive right below them. When they heard that, the team each grabbed a weapon and waited for the car to close in. Steven reached for his M16 with the M203 attached; he wanted to stop the car with a grenade. Adam grabbed an M16, as did Anthony, while Santos used the M40A3, figuring that if the car got past them, he'd try and put a shot into the back windshield.

The road was a straightaway, allowing the driver to go as fast as he could. Right when the car reached the hill, the marines opened up, but hitting the car was like trying to shoot a flying bird. Adam and Anthony unloaded two magazines apiece, while Steve lobbed grenades. Santos fired once, but the scoped rifle wasn't the weapon of choice. Unsurprisingly, the car made it through a gauntlet of fire and drove off.

When the dust cleared, the snipers were disappointed that they hadn't stopped the men. They reported the incident and headed for extract, and within the hour a MAP squad made it to the intersection. There they found two 120mm artillery rounds chained together, forming what the marines knew as a daisy-chained IED. If used correctly, it could have hit two or more vehicles at once. When the snipers learned of this, they were partly satisfied, but Santos wasn't completely satisfied until later in the week when he helped kill the insurgents emplacing the IEDs.

Four nights later, Santos and Steve were back, this time with two other snipers. Instead of the hilltop eight hundred meters away, they were now 320 meters (a thousand feet) from the intersection. When they inserted, they spent the entire night digging into a small mound of dirt, and by morning, they were buried in the mound with desert camouflage netting covering them. The team was undetectable to the naked eye.

The next day was filled with more observing. When it came time for Santos to watch, he rolled over behind the .50-cal. Two holes had been made to observe from, one for the .50, and the other for the M40A3 used by Kevin, the other team leader. Santos had only been on watch for fifteen minutes when a van neared the intersection. It slowed to cross the road, but after it did, it made a U-turn and crossed again. It turned around once more, but this time it stopped on the road.

"Get them up, we've got something here," Santos told Steve, who was monitoring the radio. The two other snipers were resting.

Within seconds the van door opened and two men climbed out. One had an AK, the other two green mines like the ones the marines had found before.

"We've got targets," said Santos.

Steve kicked the others awake and explained the situation. Their preformed plan was for the SASR and the SAW to stop the vehicle while the others dropped the men. When Santos saw the weapons and explosives, he transitioned back to the van and prepared to shoot.

The vehicle wasn't like most American-made vans, where the engine block sits in the front. It was similar to a Volkswagen bus, where the engine is between the driver and passenger. Knowing that, Santos aimed at the lower section of the driver's door; his MK-211 green-tipped Raufoss rounds would easily penetrate the door and send a secondary bullet through to the engine. In the seconds that he waited for the other snipers to prepare, Santos pulled the SASR tightly in to his shoulder; the next few minutes would seem like a lifetime.

The snipers had neither the time nor the instinct to put in hearing protection. By the time the marines were ready, the two men had made it to a hole, while the driver of the van watched. They had no clue that the snipers were little more than three football fields away.

Santos opened fire, hitting his mark. The .50-cal resounded in their small hide, causing everyone's ears to ring in pain, but Santos didn't let up. Another shot made the driver reverse at full speed, while the men outside stood surprised by the ambush.

By now Kevin put his M40A3 to use. Confused, the insurgent with the AK fired in the team's direction but with no precision; Kevin lined him up and fired a shot dead center into his chest. Steven was on the radio calling for the quick reaction force while Victor raked the van with his SAW. He and Santos caused the driver to fumble, and he drove off an embankment and slightly out of sight.

After five shots with the SASR, Santos couldn't take the pain. He used his left hand to close his ear while shooting. The others felt it as well, and Santos couldn't help but laugh when another marine let out a whimper after each of his shots.

With the van out of sight, he and Victor aimed for the man with the mines. He dropped them and ran on the road from left to right. Victor sent a burst in his direction and hit the man, causing him to face the marines. That gave Santos a perfect view, and he aimed for the man's chest, but the man staggered from Victor's bullet. When Santos squeezed the trigger, he clipped the man's side, next to his stomach, causing a gaping wound under his rib cage. From the shot, the man fell on his face and died.

The last man, the driver, was out of the van and running for his life. Luckily for him, he was on the other side of the road and only his head was exposed. Santos watched it bobbing up and down while he ran and fired a few times but missed. After a minute, the snipers saw that the man was running for another vehicle in the distance, six hundred meters away. Only the back of the car was exposed, and Santos put his crosshairs on the rear windshield behind where the driver would be. One shot shattered the rear windshield, and Santos had to change magazines. Steve launched a few inaccurate grenades, and the driver of the van limped across the road safely and got into the back passenger seat, allowing the car to speed away.

Twenty minutes later, the quick reaction force arrived. The snipers met the men at the road and inspected the damage. Kevin had shot his target in the heart, and he was barely breathing when the marines walked up. Santos kicked him in the head as he died. The other man, whom Victor and Santos had shot, lay on his face in the middle of the road. The side of his stomach was missing, from the SASR, leaving parts of his guts on the road, while 5.56mm holes from the SAW riddled his chest.

Such stories are not abnormal coming from snipers who have operated in Iraq. Since the beginning of the war, and all across the country, military snipers have earned their stay among the best and most utilized weapons among Coalition Forces, but as of July 2009 the withdrawal of U.S. troops from Iraq's major cities marked the end of fighting for most snipers.

However, the road to the finish was a hard one, and the lessons learned were paid with blood.

The combat environment in Iraq has transformed in the past seven years. In 2003, the invasion demanded speed and flexibility. Mobility kept snipers successful, but immediately following, when the war shifted to insurgency, snipers needed patience and stealth. To be effective, snipers required preparation, knowledge, patience, and timing. Beyond all else, snipers understood that a conniving enemy using unorthodox tactics awaited them on the battlefield.

Aside from weather and restrictions, terrorists groups are indeed the main enemy in Iraq. These groups make up several factions, each maintaining different ideals, and abide by certain rules of fighting unique to each group's intent. In Iraq, these groups vary from region to region, and by sects of religion. Al-Qaeda in Iraq is one of the more well-known groups, which is mostly formed by foreign fighters. Sunni and Shia groups also operate against Coalition Forces and most often keep to their local communities. Throughout Iraq, U.S. snipers have faced these fighters and have learned of their tactics.

Many of the groups have learned and utilized the same tactics against the Coalition. The most basic and effective has been the used of IEDs, ranging from vehicle-borne suicide bombers willing to kill themselves with vehicles packed with explosives, to insurgents planting bombs on the side of the roads. With patience and precision, snipers have been lethal in countering such methods.

Intimidation has also been effective for terrorists. Killings, kidnappings, mass bombings, and assassinations have threatened civilians and Iraqi soldiers and prevented them from helping Coalition Forces. For snipers, timing and preparation have been effective tools to prevent such methods.

The combat environment in Iraq also plays a part in sniper operations. Many have reported the differences and lessons learned between Afghanistan and Iraq. In Afghanistan, much of the fighting takes place in rural areas and open landscapes. There, snipers must always consider the sun's position for building hides and patrolling. Keeping to the shadows is a must, because once they are exposed, enemy contact is highly likely. Snipers have also learned that in Afghanistan, the enemy fights to keep his ground.

Iraq has posed different threats. Highly populated areas present obvious barriers for snipers. Remaining undetected can be difficult, which is why snipers move almost strictly at night. Urban areas also make it hard for snipers to hide entire teams, but many tactics can be used to counter this threat.

Overall, Iraq has shaped the way that the U.S. military prepares and uses snipers. It has been a place of hard lessons, and great successes. It is also a war zone that has presented incredible stories of snipers in combat.

SEVEN

REDEMPTION

BRAVADO and pretension aside, snipers tiptoe a fine line between life and death in the form of success and failure. Finding the enemy, or being found by the enemy, depends heavily on a few factors. Thousands of hours of training and preparation combined with extraordinary standards guide snipers toward success, but unfortunately in war, men make mistakes. In Ar Ramadi, Iraq, 2004, one incident showed that mistakes are paid for in lives. This event, as painful as it was for Marine snipers, gave a bitter lesson in death, but later it provided a remarkable example of redemption.

By the summer of 2004, combat action ribbons were guaranteed in Ramadi. The Second Battalion Fourth Marines had taken over the capital of the Al Anbar Province earlier that year and immediately traded blows with relentless insurgents and anti-Coalition fighters. For months the "Magnificent Bastards" had battled the enemy from the streets, rooftops, bases, and anywhere else the enemy dared meet them.

In that time, the battalion's snipers proved to be a hot commodity, worth their weight in gold. They had disposed of countless IED planters, repelled attacks, and survived brazen bombing. One team, though, Headhunter Two, led by resilient sniper Sergeant Santiago, cheated death repeatedly. These men braved the most firefights and hairiest engagements of all the teams in the city. Their luck soon turned, however.

By summer the team had hit a crossroad. They lived alongside Echo Company, deep inside Ramadi, at a small operating base known as Combat Outpost. Its location allowed insurgents to pester the marines with small arms fire and RPGs. In addition, that month, daily mortar attacks peppered the outpost. The marines pinpointed the enemy's firing locations with counter-mortar technology, and once a pattern was established, Sergeant Santiago's team was dispatched to eliminate the perpetrators of their grief.

Preventing IEDs on Route Michigan, the city's main road, had been their regular assignment. The new mission was a much needed change of pace, but when Santiago received the brief, he realized a problem. The command also wanted his team to resume IED prevention, meaning that unfortunately for Santiago, he needed to split his team. This was a decision he would never forget.

Though he had divided his team before, this time was different. He wanted two of his teammates for the counter-mortar mission, leaving the last member in charge of three infantrymen to make up a four-man IED prevention team. He left Tommy in charge. While Tommy was not a HOG, he had plenty of sniper operations under his belt, and he had Santiago's full confidence. The problem was that Tommy was not pleased. Normally snipers do not allow untrained men to fill the roles on a sniper team, but with a thin platoon, they had no choice in this instance. Even though the marines filling in had experience on observation missions, Tommy was left bitter about the situation.

His disappointment came with good reason. On the IED missions in Ramadi, the snipers used one rooftop the entire time and were not allowed to move positions even after being compromised. Santiago argued that this practice directly contradicted sniper doctrine, but the command's larger perspective called for the snipers to stay put because they filled a vital gap in the line of observation positions along Route Michigan. Tommy feared an inevitable attack if they continued with the same position. Nevertheless, in the Corps, mission has priority, and though Tommy didn't want to go, it had to be done.

Early on June 21, Santiago and Tommy prepared for their missions. It was Santiago's birthday, and his team shared pound cake from an MRE before leaving. Everyone enjoyed some, except Tommy. He was still sulking about the situation. Santiago wanted to cheer him up, but let him have his space.

Under darkness, the teams crept quietly through the city. Within hours they were in their respective positions. Santiago led Headhunter Two Alpha, and he monitored the radio from a building close to where the mortar men were suspected to be.

A few hundred yards away, Tommy, leading Headhunter Two Bravo, reached his building and started observation. With everyone using the same radio frequency, the teams sent radio checks to the battalion, as was the operating procedure.

Everything went as planned, but after sunrise Tommy's team had not made a radio check. Santiago noticed, as did the battalion, who tried reaching them, but with no answer. The problem could have been a number of things. No communications usually meant someone was not paying attention or that the radio had gone down. Usually it was the latter, but either way, as the team leader, it annoyed Santiago. He didn't want his snipers to get a bad reputation. He figured it to be only a matter of time before the team made contact, but when Tommy did not respond over his handheld radio, which only the snipers used, Santiago began to worry.

Within hours Echo Company diverted a nearby foot patrol to Tommy's position. Around 11:30 A.M. on June 22, Santiago and his team heard a devastating radio transmission.

"We need a corpsman here!" someone exclaimed.

"We have four marines, KIA!"

Over the radio, they learned that Tommy and his team were dead.

At first Santiago doubted his ears. The news did not seem right, but tears filled his eyes as the truth dawned on him. His teammates were shocked, wondering how it could have happened. They were unaware of the gruesome scene on Tommy's rooftop.

One marine had multiple gunshot wounds, while the others had been shot in the head; one also had had his throat cut. Tommy was found in the fetal position, wearing no body armor or boots; the marine near him lay on his side, covered in a mosquito netting to fend off the bugs. They were believed to have been resting, normal for the team members not designated on watch. The marine with the most wounds, a corporal who had survived other bloody attacks, had gunshot holes in his hands, signaling a defensive struggle. Another marine lay on his back, also appearing to have put up a fight. Everyone who knew him knew that he carried four throwing knives, and one was missing. The walls around them had bullet holes, and their gear, the radio, thermal devices, weapons, ammunition, and other equipment were missing, including the prized M40A1 sniper rifle.

The loss was unimaginable, but so were the circumstances surrounding the event. How could four marines be killed without hurting even one of their attackers? Had the marines all been asleep? Did it matter that none of them were actual school-trained snipers? These questions and more were answered after Naval Criminal Investigative Service completed a thorough investigation.

The attack was planned. Intercepted cell phone traffic indicated that the enemy had asked permission to "do these marines." Sometime after sunrise, four attackers made their way onto the roof. Santiago believes that they could have been posing as construction workers, because usually the marines placed a metal guard against the door to the roof to alert them of any invaders, whom they would have surely met with gunfire. Sometimes, though, they allowed workers to pass through to get to their site. He also suspected that the men used suppressed weapons, as not one Marine unit had heard gunshots. Santiago's team was a few hundred yards away, and Combat Outpost was eight hundred yards away.

The loss severely wounded Santiago. Though others blamed Tommy for the outcome, as the team leader, Santiago felt responsible for it all. Survivor's guilt and heartache consumed and changed him. For the rest of his time in Iraq, he wasn't completely normal, willing to die rather than let another one of his marines get injured. Even later, when a form of redemption was brought about by a fellow Marine sniper, he still would not be able to live it down.

Within the next two years grainy footage of insurgent marksmen began to surface. An enemy sniper calling himself Juba or the Baghdad Sniper posted Internet videos of his attacks on U.S. and Iraqi troops. The footage showed a series of individuals being tracked and shot by the sniper, who was not shy about his allegiance to the Islamic Army in Iraq, an extremely brutal band of terrorists responsible for beheadings and killings of journalists and civilians. The sniper acted on behalf of this group, whose main focus was combating foreigners in Iraq, especially the U.S.-led coalition.

Dark Horse Sniper

In California, Marine snipers became aware of the videos. Any chance to learn about the enemy without facing them was a treat. If this Juba was stupid enough to show his methods, the snipers would take full advantage.

One Marine sniper watching was Sergeant AJ Pasciuti. He, just like other snipers preparing for Iraq, wanted to see what he was up against. His team examined Juba's videos and noticed a few common factors.

Juba's tactics were simple and seemingly effective, but his target range was not exceptional. In all of the clips, as the sniper fired his weapon, the camera jumped, implying that he shot from a confined area. The Marine team noticed that often the camera moved away from the scene after the shooting, leading them to believe the sniper was shooting from vehicles. Pasciuti also noticed something peculiar but kept it to himself. In one scene the sniper waved a Browning high-powered .45-caliber pistol in the air. The Browning was a very uncommon weapon in Iraq, especially in the hands of an insurgent.

At the time, Pasciuti was an unlikely veteran at twenty-one years old. Normally, Sunnyvale, California, the heart of Silicon Valley and Pasciuti's hometown, produced tech-savvy software designers, different from what he wanted to become. He knew right away after 9/11 that becoming a Marine was his destiny, and by the time he could legally drink beer, he had become a product of the First Marine Division Scout/Sniper School.

Physically, he was not the most gifted, but what he lacked in strength, he made up for in heart. This was a trait that followed him through his challenges in the Marine Corps. It helped him to get past boot camp and into the Dark Horse, Third Battalion, Fifth Marines as a rifleman. In 2003, it also helped him through his first taste of combat.

During the invasion of Iraq, Pasciuti used the M203 40mm single-shot grenade launcher. When his unit crossed the border into Iraq, his young-man perception of war and glory soon faded, and the reality of death and fighting wracked his nerves.

There in Iraq, he experienced a pivotal moment in his life. He witnessed Marine snipers in action one day when his company moved to destroy a suspected terrorist training camp on their way to Baghdad. Pasciuti, watching enemy muzzle flashes from the camp, had taken cover behind a dirt mound when suddenly, a few yards away, two Marine snipers appeared and began targeting the shooter. Their weapons and equipment were different, but what intrigued the infantryman was their calm demeanor.

The two snipers easily found their target and Pasciuti never forgot what he saw next. The snipers tracked the enemy soldier while Pasciuti looked on through his four-powered ACOG (Advanced Combat Optical Gun Sight). The distance seemed astounding to Pasciuti. It was more amazing when the marine put a bullet in the soldier's head, causing him to crumble forward. The snipers' precision awed Pasciuti, and the next morning when the marines swept through the camp, he came across the dead soldier still lying there. At that moment his goal was to be as lethal and precise as the snipers he had witnessed. He wanted to become a HOG.

When the marines returned home, Pasciuti's role changed to that of company clerk. It was miserable for him, but he soon found a way out through the scout/sniper indoc. The company first sergeant initially refused his request for sniper school, but when the company gunny heard of it, he gave Pasciuti the advice of his life.

"Listen here," he said, "I'm not going to tell you what to do, but every man is responsible for his own destiny. If you're not here on Monday morning, I'll know where you're at."

Pasciuti took full advantage of the tryout. It began at 0300 on Monday morning and lasted less than a week, but in that time the numbers dwindled from forty to less than twenty. Though pushed to his physical and mental limits, Pasciuti held strong. This paid off with his acceptance into the platoon. It was an unimaginable honor, and he vowed to show that he was worthy of the selection.

Soon afterward, the sniper platoon mustered for a battalion formation. Everyone was told about the four-man sniper team killed in Ar Ramadi, Iraq. It was a somber day for the snipers. Some of them knew others snipers in the Second Battalion, Fourth Marines, a sister battalion which they often trained with. When he heard the news, Pasciuti knew right away that he never wanted anything like that announced about him.

In late 2004, his second deployment landed him in Fallujah. His team, Banshee Two, was led by Sergeant Blake Cole, and after knowing him, Pasciuti saw Sergeant Cole as one of the most complete and intelligent snipers he had ever met. Cole taught Pasciuti more about sniping than anyone else. It definitely helped that his instruction took place in the city of Fallujah, a snipers' dreamland at the time. Training covered everything from urban hides and patrolling, multiple target engagement, to enemy observation while on the job fighting real bad guys. That experience made Pasciuti successful during sniper school.

When he returned home, Pasciuti was given a shot at First Marine Division Scout/Sniper School. He had a plan; he wanted to go through the school unnoticed, but unfortunately one of the instructors took a liking to him. Anywhere else, this would have been beneficial, but for sniper school, it meant that he would become more physically fit.

By the end of the course, his heart and perseverance were rewarded. Pasciuti received the prestigious "Honor Graduate" award for the highest grade point average of the class. He was also chosen to receive the "Instructor's Choice" award, which is presented to the marine whom the instructors choose as the top all-around sniper.

After graduation, Pasciuti returned to his unit and began training for Iraq. Excitement about the next deployment buzzed among the team because of the quality of all their teammates. Sergeant Jimmy Proudman, the team leader, brought a strong leadership presence as the most seasoned sniper and had the confidence of his entire team. Pasciuti was the assistant team leader, with Scardino and Ramsey rounding out the team. The four of them were all school-trained snipers with combat experience. Pasciuti was excited to deploy as a HOG. Little did he know that all of his training and experience would help him to make history.

Counter-Sniper

In early 2006, Pasciuti returned to Iraq. His unit arrived and stayed at Camp Fallujah before moving fifteen miles south to the town of Al Amiriyah. This rural town had welcomed displaced Fallujah residents during Pasciuti's last trip to Iraq in 2004. When the marines arrived, the hold of al-Qaeda on the locals in Iraq began to slip after the people recognized the insurgents' disregard for innocent life.

Al Amiriyah is where Pasciuti formed his fondest memories of sniping. His team anxiously set out, wasting no time before running missions. Days after they arrived, though, bad news hit the battalion when two marines died from an IED; one had been a friend of Pasciuti's since boot camp. The marine took shrapnel in the neck and died instantly. It was a tough blow for everyone. The death of his friend was heart-wrenching, and Pasciuti couldn't shake the thought of it. He just hoped for a chance to spot IED layers.

That afternoon, his team set up with the infantry in an observation position close to the spot of the explosion. He was focused on the road while making small talk with Jimmy. A dust storm had formed on the horizon when, near the road, two men appeared carrying what looked to be cinder blocks. The two snipers focused in, and when one of the men began to dig, Pasciuti called the situation to higher, requesting permission to engage.

When they received clearance, both Jimmy and Pasciuti took aim. They used different sniper rifles; Pasciuti clutched the newer MK11, preferring its semi-auto ability, while Jimmy used the more traditional bolt action M40A3.

At three hundred yards, the IED culprits were not a difficult shot. Pasciuti put his crosshairs center mass on the man bending over, placing the bomb in the ground. Jimmy kept his sights on the other one. As in training, they planned to shoot simultaneously, and another marine did the honors of counting down for them.

On cue, the snipers released their poison. The MK11's recoil is nothing, allowing Pasciuti to see his target almost immediately. The bullet tore through the rib cage, and the man painfully reached for his side before falling. Unknown to the marines, the two IED layers had backup, who opened fire on them. This drew heavy return fire from the Marine snipers. Pasciuti had noticed others trying to help the two wounded men on the road, when a truck arrived and loaded the bodies in the back before disappearing. The marines later recovered an IED from the scene. This was the first engagement for the battalion and Banshee Two, but it was just one of many to come.

After a month on the south side of the town, the team decided to investigate farther north. Coalition Forces had not patrolled those areas in some time, and it was thought that insurgents moved within that area unprotested. The decision paid off as one day, while on a mission, Pasciuti found himself closer to the enemy than he would like to have been.

Two sniper teams joined together for the operation. The mission allowed the snipers to search for targets of opportunity in the unattended area. On the first day, they prepared a quality hide deep inside the thick vegetation near the Euphrates River. Ghillie suits let them blend in perfectly with the tall grass surrounding them.

In the afternoon Pasciuti and the team's corpsman held security and observation. All was silent as Pasciuti monitored the radio. He used a pocketknife to clean his fingernails while facing the corpsman. Suddenly the bushes behind the corpsman moved. Pasciuti lifted his head and noticed two men walking straight for him. One clutched a black object in his hands, but they both had AKs slung on their backs and quickly closed in on the hide.

The men were oblivious to the snipers. Pasciuti sat motionless while the men walked straight toward him. No one else in the team had seen them. Within seconds, the men were close enough to step on the corpsman's back. Just before they did, however, Pasciuti yelled, jumped up, and lunged at them with his knife.

The men were scared senseless. They had no idea what they were looking at; in their mind a bush had just screamed and was coming at them with a blade. The rest of the team reached for their weapons as the men turned and ran, followed by Pasciuti. After a few yards Pasciuti came to his senses.

"I'm taking a knife to a gunfight!" he thought.

Behind him, Jimmy and another sniper, Ramsey, stood up and aimed in with their M4s. By this time, the two men were running at a dead sprint and had split up. One went left while the other broke right.

Another Marine team member, Scardino, stood with the SAW and let loose on the man running to the left, who tried making it behind a building. Scardino's machine gun dropped him in mid stride.

Jimmy and Ramsey unleashed on the man to the right. Pasciuti watched Jimmy and Ramsey test the battalion gunner's new theory of target engagement. He had explained that shooting a target near the waist, around the pelvis area, would immobilize the individual, because that area has the largest bone mass in the human body. The gunner's theory worked. Jimmy and Ramsey left the insurgent crawling after a few shots.

The snipers inspected the bodies. Jimmy and Ramsey's target had bled to death from bullet wounds to his lower back and hips. The black object that Pasciuti had seen earlier turned out to be a video camera, a big score for the marines. From it, they learned that the men were part of a local IED cell, and the video instructed others on how to make the explosives, described techniques for burying the bombs, and gave reconnaissance of potential areas for emplacing them.

In the short months that Banshee Two lived in Amiriyah, they racked up seven kills. They stopped and revealed enemy tactics on IEDs there, but the battalion transferred to a nearby town, a place where the enemy had their own hunters.

The marines moved to Habbaniyah, a town fifty-five miles

west of Baghdad, between Ramadi and Fallujah. Establishing a presence along the main road from Fallujah, through Habbaniyah to Ramadi, became the battalion's first order of business. IEDs and ambushes reigned freely along the road. Stopping them was a challenge because of the sheer distance involved. For snipers, the mission called for their keen skill of attention to details.

At the new base, the marines learned of an enemy sniper on the loose. He was accurate and careful in his methods, and a formidable shooter. One day he struck a soldier in a guard tower just before sundown. At changeover, the soldier was about to leave but remembered his binoculars sitting on a ledge. Camouflage netting covered his position, but the binos were outside of it. As he reached from the covering, his hand was exposed, giving the enemy sniper a perfect target. The soldier's hand was mincemeat. Pasciuti knew that he had to be careful there; any slipup could cost his life.

In the following months, Pasciuti's team took to observation. They lived and shuffled through observation posts up and down the main route and helped sniff out IEDs. Their advantage was in their optics, thermals, and scoped rifles, but unfortunately, as skilled as they were, Pasciuti's team could not see into dead space. These were areas blocked by terrain or man-made objects, but the marines scattered amtracks (amphibious assault vehicles) or tanks near them, making for an impenetrable chain of surveillance—or so they thought.

It all came to a head one day after a crafty attack by insurgents. It began with an IED on an amtrack. The snipers could not help, but other marines in a nearby amtrack left their position and raced to the scene. The marines had suffered only minor injuries, and the amtrack returned to its original spot, only to be met with another IED, which caused no serious injures. The insurgents had taken advantage of the gap in the lines and planted a bomb when the second amtrack left its position. It was a good strike, but Pasciuti's team would have their revenge.

Pasciuti and Jimmy knew what to do. They pitched a bait mission to the company commander, but instead of people, the bait would be the absence of military presence. Their idea called for Pasciuti and Jimmy to take two teams while under darkness, into certain buildings. In the morning, an amtrack would show up as usual, but later it would race away as if responding to an emergency, thus appearing to leave a gap in the lines. While it was gone, the snipers would keep an eye out for any insurgents.

The captain was convinced, and the mission was approved. That night the two sniper teams, each with four infantrymen as security, executed the plan and slipped into separate buildings to cover more ground. Once set, Pasciuti broke his team up and made two sniper positions. Pasciuti's position pointed toward the main road, while in another room his assistant team leader covered another sector. Jimmy and his team did the same in their building, enabling all of the snipers to cover a tremendous amount of land.

At 0700 the next morning, the amtrack arrived, and a short time later it sped away, setting the trap. Before noon, Pasciuti took the rifle while his spotter, Doc Barth, headed downstairs to use the bathroom. Sergeant Kevin Homestead, the infantrymen's squad leader, eagerly replaced him. Kevin had worked with the snipers before and jumped at the chance to spot when Pasciuti asked him. It was not every day that infantrymen peeked into sniper operations.

Ten minutes later, another amtrack rolled into the area. Pasciuti overheard their call sign and raised them on the radio to inform them of his group's presence.

"Red One, this is Banshee Four. Just a heads-up. We're running a mission out here," he said.

"Roger that, Banshee Four. We're well aware. We're gonna hang out here for a few minutes and then we'll be on our way."

In the building, Pasciuti dropped the handset and resumed scanning the street below. Outside, an avenue ran straight from his building to the main road, where the amtrack remained. Moving nearest to farthest, he thoroughly inspected people on the streets—cars, windows, and everything else— just as sniper school had taught him. Minutes later, he came across a car. At first he did not see anything out of the ordinary and almost moved on, but suddenly, from a small window behind the rear passenger door, a glare caught his eye. He took a better look by adjusting his scope, and immediately recognized a video camera, which faced the amtrack. It dawned on him what was happening. Insurgents loved to tape their attacks for propaganda, especially IED attacks.

"Red One! Red One! You need to button up! You're being observed!" he yelled, hoping to get the marines in the amtrack out of harm's way.

"What's that, Banshee Four?" was the reply.

"Red One! Button up right now! Something is gonna happen! You're being videotaped!"

The marines in the track took cover and sealed the hatches right away. Pasciuti found the distance to the car and dialed his scope to 284 yards. Rules of engagement stated that anyone with such equipment in those circumstances fell under hostile intent, allowing the snipers to engage. Because their hide was exceptional, under the circumstances, Pasciuti did not want to give his position away and decided on other methods of engagement.

At first, Pasciuti wanted air support to destroy the car, but he had none on station. He switched to artillery and called in a mission, but nearby, the battalion commander heard the transmission and pinpointed Pasciuti's location. He denied the use of artillery because of a nearby mosque and the potential for collateral damage. Next, Pasciuti tried persuading mortars, but he was denied for the same reason. Finally he called Red One, telling them to engage the car, but they were unwilling, as well.

Meanwhile, the company commander heard everything.

"Banshee Four, take that shot!" he yelled over the radio.

"Pasciuti! Pasciuti! Look!" said Sergeant Homestead, tapping him on the shoulder. Pasciuti sighted in on the car in time to see a hand adjusting the camera. He knew that at any second, the man could strike, and with visual confirmation of someone inside, Pasciuti prepared to engage.

The car sat almost horizontal to their building. Pasciuti aimed in while Homestead spotted, using binoculars. After a deep exhalation, the sniper let the crosshairs settle just above the camera, hoping to hit anyone behind it. Knowing that shooting through glass would throw the bullet off course, his goal was to shatter the glass with the first shot and get the kill with the follow-up round.

After a countdown, Pasciuti let his bullets fly. His first shot hit six inches above the camera. The small window had an opening now and the final two shots were a tad lower. When he finished shooting, the buildings around them erupted with the sound of women wailing and crying. The neighbors must have thought that the snipers were executing the family inside the house, but within seconds all went quiet except for chatter over the radio.

By now, all units knew the situation. Pasciuti rolled off his gun to direct the track onto the car. He told Sergeant Homestead to get behind the sniper rifle just in case anything happened. Homestead was not a sniper or even in the sniper platoon, but Pasciuti needed him to hold security while he reported his actions. Just as Pasciuti began to speak, Homestead noticed a man walking up to the car.

"Pasciuti, look!" he said anxiously.

Pasciuti saw the man calmly stroll up to the driver's side door.

"If he touches that car, shoot 'im," said Pasciuti.

The man hurried to the car and opened the driver's door. Before he climbed in, he glanced into the backseat, just as Homestead fired on him. The bullet struck him below the armpit, dropping him to his knees while he looked for the shooter. Pain was written on his face, and Homestead's second shot hit him in the sternum. Surprisingly, he crawled into the car and lay on the center console. When he stopped moving, Pasciuti relieved Homestead and got back on his own weapon.

At that angle, Pasciuti did not have a great view of the man except for his legs. He put a bullet into both of the man's knees to prevent his running away.

Minutes later, the track rolled up. After surrounding the vehicle, the squad leader approached it, while Pasciuti spoke to him over the radio, guiding him to the camera in the backseat. Pasciuti watched while the marine waved the camera in the air.

Blood filled the cab of the car. The man in the backseat had died almost instantly, while his friend took a few minutes to pass on. The marine inspecting noticed something else.

"Hey, there's a weapon in here," said the squad leader. He handed the rifle to another marine, who almost put it in the track, until Pasciuti noticed the butt stock. Once he saw it, he froze.

"Wait, Red One. Show me that weapon," he said over the radio.

Any Marine sniper would have recognized it. The rifle's stock looked exactly like the McMillan fiberglass stock of the M40A1 made specifically for Marine scout/snipers. The marine held it in the air, and Pasciuti asked him a question to be sure.

"Red One, look for a serial number on the barrel near the stock. Does it begin with E676?" said Pasciuti.

"Roger, that is affirmative. It reads E676. Remington 700," was the reply.

Pasciuti was shocked and overjoyed. Killing an enemy sniper was one thing, but one using a Marine sniper's rifle was even more gratifying. From that, he realized how the enemy employed their sniper teams. The insurgent's sniper tactics called for the spotter to exit the car to draw less attention to the vehicle; then the sniper would shoot, and record the action. After hearing the shots, the spotter would then get back in and drive the vehicle away. It was a clever method until then.

Later, the marines pulled the car to base for inspection.

They discovered a hidden compartment in the trunk, revealing grenades, U.S. machine gun ammunition, fake passports, and a Browning high-powered .45-cal pistol, exactly like the one Pasciuti noticed in the Juba video. From the bodies, IDs were recovered, leaving Pasciuti with an impression about the shooter's background. He was clean-shaven, nicely dressed, with no prior record, a real professional.

Days later, Pasciuti shot the recovered rifle. The Unertl scope used by Marine snipers was missing, and Pasciuti understood that the last marine using the rifle must have done his job and destroyed it. Instead, a crude Tasco 3–9 power scope was mounted, but after firing the gun, Pasciuti felt that any shooter could have gotten the job done with it. At one hundred yards the bullet hit half a minute right, or half an inch to the right. Insurgent snipers hardly shot farther than that, and by default, the shooter must have been deadly with it. Pasciuti wondered if the man he killed could have been Juba.

When Pasciuti's deployment was finished, he learned that the rifle belonged to Tommy Parker from the Second Battalion, Fourth Marines, the team leader of the fallen sniper team in Ramadi in 2004. Pasciuti saved the chambered bullet, and with the rifle, he made a plaque and presented it to the Fifth Marine regiment headquarters.

The contrast of the success and failure of the two teams is a strong lesson for any sniper. After Tommy's incident, many sniper teams were forced to operate in no less than six-man teams, making for a bigger footprint and a greater chance of compromise. Regardless of fault, lessons can be learned from this tragic incident—namely that snipers thrive when allowed to think on their own. They should be employed according to their skill set and should never be forced to operate against sniper doctrine.

Pasciuti's success is a prime example of what snipers are capable of when they are allowed to think and operate in their primary role as hunters.

EIGHT

SUNNI TRIANGLE

IT is an eerie feeling knowing that someone is trying to kill you. In Iraq's deadly Sunni Triangle, that was exactly what American forces felt daily. Starting above the northern city of Tikrit, the triangle spans southeast past Baghdad and westward to Ar Ramadi. It was the center of support for the former president Saddam Hussein, and when he was ousted, it became the center for Sunni opposition against the U.S.-led coalition. Living there for troops was about as comforting as sleeping with an M4 under your arm. From the first day he set foot in the triangle, U.S. Army Sergeant Adam Peeples was, just like every other American soldier, wanted dead by the enemy. After two deployments there, he was blessed to make it out with his life and limbs.

At eighteen years old, young Peeples left Griffin, Georgia,

to join the Army. In high school he had never considered the service, but he figured that if he wanted to marry his high school sweetheart, the Army would provide some stability. When the time came to sign up, he wanted a challenge and the occupational specialty of 18X. Special Forces was just that. During the physical, however, he was not able to pressurize his ears, an essential physical quality for being airborne. Since Green Berets are airborne, it meant that Peeples was bound for the infantry and that he would have to be a leg, meaning non-airborne.

In 2003, Peeples became an 11B infantryman. After boot camp, his assignment was to Bravo Company, First Battalion, Twenty-sixth Infantry of the First Infantry Division based in Germany. Peeples arrived just as the war in Iraq kicked off, and because his unit was mechanized infantry, he was trained to drive Bradley Fighting Vehicles. Though he was not too thrilled about driving, the comfort of a light-armored vehicle equipped with a 25mm cannon in an urban environment was reassuring. After all, where he was going, every advantage would help.

Round One

His unit, the Blue Spaders, arrived in Iraq on Valentine's Day 2004. They were not welcomed with flowers and chocolates, but instead by a tenacious insurgency at the center of the Sunni Triangle, Samarra. As one of Iraq's four Islamic holy cities, Samarra encompassed ancient history and precious monuments, but Peeples cared nothing about that. He cared nothing about the ancient ruins, the golden mosque, or any other tourist attractions. All he knew was that Americans were not welcomed in Samarra, and that before his arriving there, insurgents had united to resist the American-led coalition. Their network and communications stretched throughout the city, resulting in unpredictable attacks on U.S. patrols and bases.

In Samarra, Peeples's unit lived outside of the city. It occupied a small, desolate outpost known as FOB Brassfield-Mora, named after two soldiers killed the year prior. The base had no showers, no chow hall, and only dust and sand for amenities. Peeples was amazed at the lack of cultivation in the area. The Iraqis were years behind. That aside, the soldiers quickly learned that the enemy were capable of urban tactics, and they were thrust into the thick of things. For Peeples, training had long since passed. Now he would experience the realities of combat.

Soon his platoon was the designated security detail for the local ODA team. When the Special Forces soldiers needed their assistance, Peeples's platoon rushed to their aid. In the meantime, however, their lives consisted of patrolling roads and providing escorts. The Special Forces Green Berets operated from a small compound made up of two adjoining safe houses on the western edge of Samarra. Their location allowed the soldiers to infiltrate the city quickly and return safety in the same manner, but the convenience came with a price. Being that close to the city gave insurgents the freedom to strike their base with ease.

One day, a massive group of insurgents assaulted the compound. Peeples's platoon responded, and the soldiers took to their vehicles hooting and hollering as they headed for the fight. For many it would be the first time under fire, and they would finally get to see what they were made of.

Peeples was anxious during the drive to the compound. At the scene, the Bradleys positioned themselves for maximum security. The gun battle was raging when Peeples and other dismounts moved onto the rooftops under direction from the Green Berets. When he left the Bradley, Peeples managed to swipe an M240 machine gun. On the roof, he surveyed the neighborhood, ready to meet his first target.

The area looked like any other Iraqi city. Beige buildings blended for entire city blocks, with a few mosques with minarets scattered about. Countless telephone wires linked homes and structures, while run-down cars lined the streets. A cemetery sat in front of the compound and stretched to 100 meters (330 feet) away at the base of a wall. Beyond the wall was where the insurgents occupied several five- to eight-story buildings.

At first glance, Peeples had a hard time differentiating the tan structures. They seemed to melt together. That soon passed, however, and ten minutes later, he spotted an insurgent. The gunman fired from a window, aiming toward the adjacent compound. Without hesitation Peeples threw a hundred-round burst at the man's position. With bullets landing near him, the insurgent ducked for cover and immediately others opened up. It was the beginning to an eleven-hour firefight.

When Peeples looked up, he saw insurgents firing RPGs and machine guns. A massive volley of rocket-propelled grenades exploded near the soldiers. Some destroyed pieces of Peeples's building and the headstones in the cemetery below him. Bullets began to hit the wall in front of Peeples, throwing chunks of cement in his face. Instinct caused him to react. He unleashed with his M240 machine gun until another soldier arrived and took over, forcing Peeples to use his M4. The two soldiers were alone on the roof and held out as long as they could, but they were severely exposed with little cover, allowing insurgents to put rounds inches away from them. The fire was so close that Peeples pulled the back of his vest into his helmet so that he would not be shot in the back of the neck. At the first lull in fire, the two retreated from the roof, but they were ordered back.

Though shaken, Peeples resumed his position. His M68 red-dot reticle allowed for easy use; anywhere the red dot was, the bullet would hit. At first he shot at muzzle flashes from windows, but he did not seem to hit anyone, so he changed his tactics. Hours into the fight he learned a critical lesson in patience, which paid off.

He had decided not to shoot at everything. Instead, he let the enemy shoot and move, hoping they would get comfortable and make mistakes. Soon other soldiers took to his roof, and he and a SAW gunner moved to the back corner of the building to cover their flank. There, his plan worked.

From his position, he caught two insurgents moving toward them at 150 meters. In head wraps and sweatpants, one fighter held an AK, while the second carried an RPG with a bundle of rockets on his back. He struggled to keep up with the first one because he was fat, which slightly amused Peeples. He aimed at his large stomach.

It took five bullets to drop him. His stomach jiggled and he squirmed after being hit. The man next to him stopped behind cover when he realized what was happening. Peeples shifted to him, but the man had ducked around a wall. Moments later, however, his head turned the corner to look toward Peeples, trying to find the origin of the shooting. He did not leave enough time for Peeples to aim in on him, but Peeples rested his red dot reticle where the man's head had appeared, and he waited. When the man turned the corner again, a slight squeeze of the trigger put a 5.56mm bullet in his face, and his body fell motionless.

Looking at the dead in the streets, Peeples had a realization.

On the ground, Bradleys bombarded positions with their 25mms, and overhead F-15s and F-16s along with AC-130 Spectre gunships put scunion on the enemy. When he signed up for the Army, Peeples had known that war was possible, but now that it was at hand, it felt surreal. He could not stop thinking about the two insurgents he had killed. Now, suddenly, just as quickly as it had started, the fighting was over.

In the end, Bradleys were the deciding factor. On a few occasions, Peeples witnessed their impressive capabilities. Once, two Bradleys converged on and virtually destroyed an entire building when a gunman unloaded from it. Another time an RPG gunner tried downing one of the fixed wings that was in the act of a gun run. The gunner stood atop a high-rise and fired, allowing three Bradleys to lock onto his position. All at once they opened fire, decimating most of the gunner's body. Peeples watched some of his flesh fall from the building. Other times, insurgents stuck their gun barrels through small holes in the compound's walls, but with thermal sights, the Bradleys easily found them and switched to armor-piercing rounds, punching through the walls and killing them. The insurgents who had scaled the wall itself and low-crawled toward the soldiers in the cemetery had no chance.

The next morning Peeples and his platoon were back at base. Everyone reminisced about their first fight, discussing in finer detail how the events had unfolded. Their platoon was the talk of the battalion, as it was the unit's first major engagement where the dismounts were able to fire. Though he had done well himself, it was the actions of the battalion's snipers that left a lasting impression on Peeples.

During the fighting, Peeples kept an eye on the snipers. It was not until after nightfall that he got an up close and personal experience of sniper operations. On the roof next to him, the shooter handled the Barrett M107 .50-cal sniper rifle with a thermal sight attached. His spotter spied the area with night vision while Peeples listened and watched how they went about finding targets. When it was time for the snipers to act, their approach was precise and professional

"I've got two guys," explained the shooter, quickly directing his spotter onto them.

"Roger that. Weapons confirmed," agreed the spotter, and he relayed the distance.

The snipers were calm and collected. Peeples had never seen such a casual approach to killing. It was strictly business-like. He hurried to find their target, hoping to witness the skill of the snipers. The heavy rifle boomed and a man dropped at eight hundred meters (half a mile) away. Peeples knew right then that he wanted to become a sniper, but it did not happen on that deployment. He still had a long way to go, and would live through much fiercer fighting.

By fall of 2004, the insurgency controlled Samarra. The city's supposed governing body, the city council, was involved with peace talks, but the committee held no real power. The true controllers were foreign and domestic anti-Iraqi fighters along with several crime families and tribes. They simultaneously competed for power and joined forces to attack U.S. and Iraqi authorities. By the end of summer, insurgents had complete control. Their attacks plagued the U.S. soldiers so much that even patrols into the city were out of the question. It was just too dangerous.

One final attack quickly changed the insurgents' position and prompted the Iraqis and the Coalition Forces to react. It started with a suicide bomber dressed as an Iraqi police officer in a truck packed with explosives. He detonated the bomb at the Iraqi Army compound, which started a small arms fight. The attack left five U.S. soldiers dead and wounded others. The event hit Peeples hard. One of his close friends had died in the attack. To add insult to injury, when it was finished, many of the Iraqi Police and National Guardsmen abandoned the post that Peeples's friend had died protecting. The post also happened to be the last Coalition foothold in the city.

In October, U.S. forces were left with no choice but to execute Operation Baton Rouge. The planning had taken months, with small intelligence-gathering missions into the city giving U.S. forces an idea of the enemy fighters and expected responses. With enough intelligence collected, U.S. forces commenced the operation. The mission called for U.S. troops to drive the insurgency from Samarra and regain control for Iraqi forces.

The operation bred a mix of intense moments for Peeples.

The night of the attack, his unit led the way into Samarra. He sat crammed with seven other soldiers in the back of a Bradley with an M240 machine gun, extra ammo, his pack, and a radio all stuffed between his legs. They pushed in from the west and immediately met a well-prepared enemy.

In Samarra, five hundred to a thousand fighters waited. Most of them operated in small teams and used the cover of darkness to move about their preplanned positions. Command cells called for many of the teams to meet the Bradleys at the main bridge leading into the city. As soon as the enemy fighters heard the vehicles, they opened fire. Inside his Bradley, Peeples heard the whistle of falling mortars and the unmistakable sound of exploding IEDs. Bullets from enemy machine guns and AKs ricocheted off the vehicles, clarifying for the soldiers exactly what they were up against.

Peeples was the radio operator and stuck close to his platoon sergeant. Throughout the fighting, he watched how the senior soldier held up under pressure. He would also learn a lot about himself in the days to come.

Explosions filled the first day. Insurgents held many fortified positions and stopped the soldiers' advance—that is until air assets and tanks moved in. It was an impressive show from then on, with images of entire buildings being destroyed. Yet the most vivid image on that first day was a grenade from an M203 grenadier decimating an insurgent only fifty feet away. The man's body disappeared when the grenade hit him.

On the second day, Peeples's platoon took a building across from the Al-Askari Mosque. Its golden dome appeared out of place. Under other circumstances, Peeples might have loved to observe the one-thousand-plus-year-old structure, but just then, all he could focus on was survival.

At the time, his platoon remained the deepest in the city. They holed up in a hotel overlooking the main roads. Fatigue wore on them, and it showed when Peeples and a few other soldiers initiated an ambush. They planned to simultaneously shoot a group of men, and they were to open fire after a count of three. The soldiers were so tired, however, that they had forgotten to take their weapons off safety, and after the count, no bullets were fired.

Later that day, Peeples spotted men carrying RPGs and small arms being loaded inside a car as it sped around a corner and into the sights of a Bradley. The fighting vehicle wasted no time in annihilating the entire car. Men from the car tried climbing out, but they were all killed. When the shooting was finished, bodies lay in the street while the car smoldered. Sometime later, Peeples saw an old lady kick one of the bodies while walking by. Shortly afterward, a dog chewed off the leg of one of the men and dragged it away.

Day three marked the end of the fighting. From there, Peeples and his unit took over a building where they stayed for the rest of his deployment. His platoon made friends with a local family there, and the soldiers realized that the civilians liked them. This one family even cooked for them when they came around. Their interaction was bittersweet, and in the end, Peeples learned that the family's fourteen-year-old son had been beheaded for helping the U.S. soldiers.

Through it all, Peeples learned a great deal. His personal confidence soared after having survived such heavy fighting, and he felt that he could handle any situation. The biggest lesson he learned, however, was to stay calm and react rationally under fire. That mentality helped him make it to his battalion's sniper section.

Peeples returned to Germany in May 2005. After a month of leave, he was back in action and volunteered for the battalion's week-long sniper selection. The summer heat had an effect on the soldiers, and from the beginning, Peeples watched others quit or drop and vowed not to be one of them. One day, as if it was not hot enough, after finishing a two-mile stretcher carry, the candidates were put through sniper disciplines and were made to lie in the red clay dirt in a simulated sniper position. They were not allowed to move for hours, testing their bodies and their minds. When they finally finished, two soldiers were dismissed for moving.

By week's end a total of three soldiers, including Peeples, had survived. The other thirty-something were dropped, or quit, or could not finish the final event—a nine-mile run with an eighty-pound pack. As gratifying as getting in was, Peeples did not have time to celebrate too much. Though he was part of the platoon, he was not a sniper yet, not until he passed sniper school.

Within months, Peeples was crawling his way through the brigade's pre-sniper course. There he learned the basics of sniping in shooting, stalking, and mission planning. Though it was a precursor, he took it seriously. The course taught him the essentials, which he needed, because immediately afterward, he was sent to the U.S. Army National Guard Sniper School in Arkansas. Everything that Peeples had learned from the pre-training was developed in finer details there. It made it easier for Peeples, and he passed the course, becoming a certified B4, a U.S. Army sniper.

By now Peeples enjoyed the Army. His family helped him to grow, and they were comfortable with his being in the military. He had survived his first deployment and picked up the rank of sergeant. Back in Germany, with the other snipers being discharged, Peeples slid into a team leader billet. It all seemed to fit, and he figured that he might re-enlist, depending on this next deployment.

Within months, the soldiers were training for Iraq. Once Peeples was acquainted with his new teammate, Specialist Stout, they were joined at the hip. Stout was reliable and hardworking, and though he had not attended sniper school because of an injury, Peeples trusted him completely. The two of them worked on their standard operating procedures and learned each other's habits.

They also invested in personalized M4s. Peeples bought an Olympic Arms ultra-match twenty-inch barrel with a free-floating tube, which made his standard M4 more accurate. He added a JP adjustable gas system that controlled the rifle's recoil, allowing him to acquire targets quickly after firing. He also had a few other modifications done and was allowed to use the weapon after combining his personalized upper receiver with a standard-issue M4 lower receiver. Along with his M24 and a .50-cal, he and Stout had a nice selection of weapons, which they were about to put to use in Iraq.

Round Two

The soldiers knew that the sandbox was their destination, but the exact city was unknown. At the same time, one of Iraq's most dangerous cities, Ar Ramadi, made daily headlines. Images of suicide bombings and face-covered insurgents waving AKs and declaring war on the U.S. demonstrated the type of enemy there. Peeples sarcastically boasted that they were going there; at least he wanted to, knowing the city was action-packed. He did not want his next deployment to be boring, and amazingly as the soldiers readied to deploy, Peeples's wish came true.

As mechanized infantry in the Bradley fighting vehicles, the 1/26 was ordered to send a company to Ar Ramadi. There, they would support the existing battalion, which had been in heavy fighting, while the rest of the 1/26 headed to Baghdad. By chance, Peeples and Stout were chosen to support the company in Ramadi, and they could not have been happier.

The soldiers flew into Kuwait and drove into Iraq in 2006. Their convoy took less than a week, and during that ride the soldiers only talked about Ramadi. Its reputation was that all U.S. troops there were in for a fight. Some of his peers seemed fearful, but Peeples could not blame them. The damage of IEDs, snipers, and attacks had taken a toll on U.S. troops. Plus the Bradleys were guaranteed to be in the thick of it. Regardless of the danger, Peeples wanted to be there.

In Ramadi, the soldiers arrived at Camp Corregidor. The outpost stood surrounded in the center of the city, and there Peeples's unit met their new battalion. They also met Charlie Company, 1/6 Infantry, the other mechanized infantry unit that they were replacing. Right away, Peeples made contact with their men to learn as much as he could about the Area of Operations (AO). The soldiers complained that they had not seen much action and that they had only had one engagement. That did not sit right with Peeples, knowing that the city was full of insurgents.

He found the battalion snipers and talked with them. After asking them the same questions, he was taken to their living quarters and shown the wall outside of it. The wall was riddled with bullet holes.

"That's what we've been doing," said one sniper, signaling that they had been trading fire with insurgents. The snipers had also been keeping track of their kills on a wall with tick marks from a black Sharpie marker. The tally was well over a hundred.

"There's no shortage of action here," the sniper said.

The snipers spent hours explaining Ramadi to Peeples. He learned from where terrorists operated, along with their tactics and preferred methods of attack. The snipers explained that he should expect enemy sniper fire, small arms, and mortar attacks a few times a week, if not daily. The city was the enemy fighters' territory, and they had become cunning in their ways of defending it, especially in using the deadly IEDs. As a sniper, Peeples would be expected to sniff out these attacks and protect the vulnerable foot patrols. However, as a sniper he would also be targeted more so than any others.

He also learned about who the enemy were. Some were local tribesmen with allegiances to different sheiks. They did not agree with the Iraqi government's cooperation with the U.S. and worked to undermine any progress. Others were al-Qaeda operatives and foreign fighters; they threatened violence to Iraqi civilians and soldiers if they dared help the Coalition. Their brutal intimidation made it tough to bring peace to the city. Just as one of the soldiers had read on a wall, insurgents considered Ramadi the graveyard of the Americans.

Once the relief was final, the company began operations.

Ian Baker and partner
somewhere near
Kandahar, Afghanistan.

COLLECTION OF IAN BAKER

Ian Baker and team.

COLLECTION OF IAN BAKER

Ian Baker and team
in Pakistan.

COLLECTION OF IAN BAKER

Vechicle after ambush involving Ian Baker.

COLLECTION OF IAN BAKER

Bobby and his SF during extract.   COLLECTION OF BOBBY MICHAELS

Bobby in Afghanistan.

COLLECTION OF BOBBY MICHAELS

Bobby in GMV during a convoy.   COLLECTION OF BOBBY MICHAELS

Bobby on the hunt.   COLLECTION OF BOBBY MICHAELS

Chris and his Navy SEAL sniper team. COLLECTION OF CHRIS OSMAN

Stan in training.
COLLECTION OF STAN CROWDER

Stan in Afghanistan.
COLLECTION OF STAN CROWDER

AJ and Sergeant
Homestead after
recovering a fallen
Marine sniper
team rifle.

AJ and the recovered
Marine sniper rifle.

AJ protecting his base.
COLLECTION OF AJ PASCIUTI

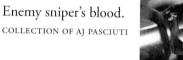

Enemy sniper's blood.
COLLECTION OF AJ PASCIUTI

View from AJ's hide.   COLLECTION OF AJ PASCIUTI

Peeples and his sniper
partner in Ramadi.

COLLECTION OF ADAM PEEPLES

Peeples during second deployment. COLLECTION OF ADAM PEEPLES

Peeples on target.

Peeples with periscope.

Johnny in a hide.

From the start, Peeples found it difficult to get sniper missions. His commander hesitated to use them and worried about their safety. It took time and patience for Peeples to convince him that his small team could handle themselves as long as they had support.

In the meantime, Peeples's first action as a sniper came nights later. He and Stout were manning observation posts to prevent ambushes or the planting of IEDs. One night, just as he had attached the PVS-22 universal night sight to his scope, Peeples's camp came under small arms fire. His night vision was great with the city lights, which brought a very clear visual for hundreds of meters in each direction. He scanned rooftops for muzzle flash and instantly found a target. It was an insurgent shooting at another observation post, oblivious that Peeples was aiming at him.

With four hundred meters (a quarter mile) between them, Peeples aimed for his chest and let two bullets fly. The man was hit and went down, but he sat up again with his weapon in hand. Peeples aimed once more. His first bullets had been on point and he did not need to adjust his sights. The man stopped moving when Peeples put two more bullets in him.

Peeples kept his aim on the rooftop. He knew that others might appear, and moments later someone did. A man emerged with his hands on his head and walked up to the dead insurgent. Peeples did not shoot. He watched to see if the man would pick up the AK, but he kicked it away and carried the dead man downstairs. Peeples let him live. Though this was not his first kill, or even his first firefight, it was his first time behind a sniper rifle, and having that technological edge made all the difference.

Two nights later, Peeples sat eating in the command post. The camp had been under fire lately, and he kept his team ready to react. Suddenly, over the radio, an observation post requested a sniper. Through their thermals, the soldiers on post had noticed a man observing them from a window. Peeples grabbed his rifle and ran to the post across the street. He felt a little spooked about running alone through the streets of Ramadi in the middle of the night. Once he reached the building, the other soldiers let him in.

Inside, he confirmed the target. That room, however, did not provide a great view, and Peeples decided to climb onto the roof for a better angle. The added height helped, and this time when he sighted in, he saw through the target's window. There, a man was in the process of loading an RPK machine gun on his lap.

Peeples readied his weapon. He guided the soldiers below him onto the target, asking the senior soldier to spot for him. The soldier used thermal vision gear and excitedly agreed. Peeples found the range and fixed his scope. When he was ready, he took aim. The bipods steadied the rifle, and his butt stock rested comfortably in his shoulder. The crosshairs lowered onto the man's chest. After a deep breath, Peeples lightly squeezed the trigger. His bullet, however, hit a power line between them and sent sparks into the air. Quickly, Peeples scooted right and found his target again. The man had noticed the sparks but did not realize that he was being shot at and did not move. Peeples wasted no time. He chambered another round and took one more shot.

"Wow! Blood splatter from his chest!" yelled the soldier who had been spotting.

That kill gave Peeples's team full confidence from his commander. It was still early in the deployment, and the officer realized the effectiveness of snipers. He allowed them the independence to be successful. Also, when the platoon of Navy SEALS who had been living at the same compound asked for support, he let Peeples and Stout operate with them.

The frogmen, whom the soldiers referred to as NSW, or Naval Special Warfare, ran high-intensity missions. They lived, trained, and worked side by side with Iraqi Army soldiers and took them wherever they went as regulated by their mission. The SEALs were a laid back and tight-knit group, but when it boiled down to the business of war, they were professionals. From his first mission with them, Peeples enjoyed operating alongside of them.

One afternoon, a SEAL sniper met with Peeples. The platoon requested that Peeples's team support them while apprehending a high-value target. The target, a local man forging fake IDs for foreign fighters, made a living selling the IDs, which allowed the foreigners to move about the country illegally. The SEALs were tasked with taking him down. The plan called for Peeples's team and two other sniper teams, made up of SEALs, to move into the man's neighborhood and set up in a triangle perimeter around the target's house. When he showed, the SEALs would raid the house.

The next day, the plan unfolded. Peeples and Stout slipped into a building and got eyes on the objective. The SEALs set up in their hides, as well, but after sunrise one of their teams was compromised by children. The kids threw rocks onto their rooftop but were scared away with flash bangs. Shortly afterward, the target showed and the SEALs took him with no shots fired. Though the plan was a success this time, the SEALs had been fortunate to be compromised by children, because had it been insurgents discovering them, Peeples knew that the outcome would have been worse.

In time, the company was not able to provide Peeples with a security detail or a quick reaction force. The patrols needed every soldier available, which worked for Peeples, allowing his team to work exclusively with the SEALs. They had fun missions anyway.

Peeples quickly made friends with them. They were easygoing and surprisingly took his advice; he had expected judging by their reputation, being Special-Ops, that they might overlook his views. He also admired that they were not scared of action. They craved contact with the enemy and even drew up a mission to infiltrate the nastiest, most vicious area of Ramadi, Papa 10. Insurgents controlled that area and were known to use the neighborhood as a staging point for attacks. No units dared enter, but the SEALs, along with Iraqi Army soldiers and Peeples's team, drew up a simple plan with a simple objective: move in to that area and pick a fight.

Less than a week later, the men were in Papa 10. They held three separate positions inside homes, while Iraqi Army soldiers guarded the families. Each group was armed to the teeth with machine guns, sniper rifles, grenade launchers, and other weapons. Their entire purpose was to ambush insurgents.

Peeples's element held OP-3. The SEALs' platoon chief took command there and allowed Scott, a SEAL sniper, and Peeples to discuss the best method of employment. Peeples advised against the roof, but that is where Scott wanted to be.

"We don't need to take the roof. That's the last place we want to be. If we get on the roof, we're gonna suck up a grenade," said Peeples. If they were compromised, he knew exactly how the insurgents would react. "We'll still be able to see from inside the house," Peeples explained.

Scott wanted the maximum vantage point, which the roof allowed, but: "OK," responded Scott, "I trust you on this."

Even the SEAL platoon chief took Peeples's advice, to stay off the roof.

When they finished talking, they moved to different positions inside. Scott took his MK-11 and set up on the stairs, looking through a window but far enough back to where he was unnoticeable. Peeples took up a top-story bedroom. Inside, thick curtains covered the windows, blacking out the entire room. With his knife, he made two small slits in the curtains so that he could see through.

The other teams, OP-1 and OP-2, had plans of their own. They preferred the maximum vantage point of the rooftops, and by mid-morning all three positions were ready and on the lookout for insurgents.

Peeples and Scott had interlocking views. All was quiet at first, and hours passed before they both spotted an older man acting suspiciously. He had been walking from his house, only thirty yards from Peeples, to one of the other observation positions. He looked to be planning something, and finally he stepped from his house with an object under his arm. Scott lost sight of the man and could not distinguish the item. Peeples, though, still had him in view. He saw that it was an IED, but before he could shoot, the man disappeared. Minutes later he returned, but moved too quickly for Peeples to shoot. Peeples alerted Scott.

"Scott, the man's about to come back into your view. He had an IED. Take him out," Peeples relayed.

Scott's rifle had a suppressor and he took aim. Peeples reminded him to open the window in front of him before he fired, but Scott wanted to shoot through it. When the man appeared, Scott fired twice, but both shots were deflected by the glass and missed. Peeples watched the man run to his house unscathed.

In his courtyard, the man trembled. He had heard the shots but did not know where they had come from. Peeples knew he had to kill him and he cracked his window. By now the man was only thirty yards away. Peeples's picture of him was clear. The enemy's eyes were wide, and sweat gathered on his forehead. His wife met him outside, but Peeples could not let him get away, even with her there. He looked the man in his face one last time and squeezed the trigger. Just as he shot, though, the rifle slipped from his shoulder, sending the bullet over the target's head.

"I can't believe I missed," he thought, knowing that he had done the impossible by missing a thirty-yard shot.

The man ducked behind his wife. The two of them were even more scared but still had no clue where the shooting had come from. By now the wife was frantic, and the man held onto her legs, using her as a shield. After a minute, though, he stood back up, feeling certain that the danger had passed. When he did, Peeples put the crosshairs between his eyes and shot.

The bullet hit its mark. The man's head split open and brain matter splattered on his wife, who immediately went berserk. She wailed at the top of her lungs, drawing the attention of another man who dragged the body into the house.

After the shooting, the snipers decided to stay in place.

They would rather not expose themselves while trying to take up other positions, plus the area went quiet again. Minutes later, while Scott and Peeples discussed the event, explosions shattered the silence.

Gunfire followed, coming from OP-1's position. From Peeples's window, he could not see anything, but the platoon chief, monitoring the radio, learned that OP-1 had been compromised and had taken grenades on their roof, injuring a few of the SEALs. They needed a casualty evacuation. Minutes later, they were hit again.

"We need to back them up," said the chief, ordering his men to break down and prepare to move to the other team's position.

The team began packing. Everyone knew the danger ahead. Where they were, running through the streets in the open would have normally been suicidal, especially with a small team, but no one gave it a second thought. Peeples mentally prepared himself for what was at hand. At the same time, he held security while everyone gathered the equipment.

He watched the street in front of the house, leading to OP-1. The narrow road was lined with cars and courtyard gates on both sides. Gunfire came from that direction, but Peeples still did not see anyone. Suddenly five insurgents rounded a corner at 150 yards away. They reloaded their weapons and seemed to be forming a plan.

"I've got bad guys over here!" Peeples shouted.

Two SEALs rushed to his position. The heavy weapons operator held the MK-48, a light machine gun. With a look of pleasure, he, Peeples, and the chief prepared and on cue opened fire, downing four of the men. One rolled on the ground in pain. Immediately afterward, the entire team was ready to move and ran through the front door into the street, making their way to the injured SEAL team.

Once they were outside, the events became a blur for Peeples. A serious gun battle began, and insurgents gathered to press Peeples and the others. Some of the fighters took to the high ground and shot from windows and rooftops. On the street, they threw RPGs and hand grenades. Though Peeples and the others were heavily outnumbered, their tactics and superior firepower evened the fight.

The SEALs and Peeples went into kill mode. The heavy weapons operator and another SEAL started it off and let loose a wall of bullets forcing insurgents to take cover, while the rest of the team bounded. Peeples took cover behind a car. He waited for the others to follow before opening fire. When they did, he and a SEAL gave the insurgents hell.

Peeples lined his targets up and squeezed off two shots. He made it a point to fire only a few rounds per target, knowing that he might have to conserve ammo. Beside him, the SEALs sprayed every target; they executed their standard operating procedure for breaking contact with the enemy, bounding, and moving away. It was precise and exact.

Though the work was hot and exhausting, neither Peeples nor the others stopped. Once, Peeples found a man in a window with a machine gun, his face locked in anger, as he threw bullets into the wall next to Peeples. Peeples returned fire, knocking him down. As he and the rest of the team bounded, explosions and rifle discharges echoed off the concrete. Peeples even noticed a hand grenade fly over his head, exploding in an adjacent courtyard.

In less than five minutes, the team was at OP-1's house. Two Bradleys were there already. They were the Quick Reaction Force. Peeples's team waited for the OP-2 team to arrive before moving into the house of OP-1. When the other men finally arrived, Peeples's team went in and witnessed the carnage.

The road in front of OP-1's house was filled with blood. Two separate blood trails led from the dirty street into the courtyard and on into the building. Peeples could not help but wonder what had happened. In the house, pools of blood formed on the floor, and the other SEALs were distraught. Peeples overheard one of the men explain to the chief what had happened.

As a result of taking the rooftop, the team had been compromised early on. They had also been set up by the old man that Peeples had killed, who had set a pressure-plated IED at the front gate of OP-1. Later, someone tossed a grenade on the roof, hitting a few SEALs, but nothing serious resulted. The team leader called for the quick reaction force and a medivac. When the vehicles arrived, the men ran from the house, straight out of the front gate. The two Iraqi Army soldiers who led the way were instantly vaporized upon triggering the IED by stepping on it. Behind them, two SEALs were severely injured. One had 90 percent burns on his body and lost a leg, while the other had both legs broken.

The sight disheartened Peeples. The lives of both men would never be the same. It just as likely could have happened to him. While sitting there, he had an unexpected thought. He asked if anyone was guarding the roof. Nobody was. He immediately ran up there with a SEAL in trace.

Before walking out, Peeples slowly opened the door to the roof. Fifteen feet away, two men crept toward the OP-1 house from the next building. Weapons in hand, they wore green vests and face wraps. They had not seen Peeples. Peeples lifted his weapon and pumped two bullets into each of them. Moments later, more insurgents fired on the Peeples team, forcing them to take cover. Peeples and the SEAL returned fire but were ordered off the roof, because the Bradleys were going to hose everything. When the Bradleys stopped firing, the area went quiet once more.

As the fighting finished, one thought haunted Peeples. He could not help but picture the faces of the family in the house of OP-1. They were absolutely terrified, and Peeples could see it in the children's eyes. It dawned on him that everyone there would feel the effects of war for the rest of their lives.

After the fight, Peeples's team supported the company again. They had done a few counter-IED missions and held security in the observation posts before another large mission arose. It came in Operation Murfreesboro.

For some time, al-Qaeda operatives had moved freely through the eastern Malaab District of Ramadi. With support, or fear, from civilians, the district was a safe haven for them and other foreign fighters. From that area, the insurgents planned and executed many of their operations. It was also suspected that they kept many weapons caches and IED factories there. To rid them of the area, U.S. commanders drew up a plan to move in, set up concrete barriers around the Malaab District, and sweep through, tightening the noose around and isolating insurgents there.

Peeples's team began to prepare for another fight. Soon he learned that the operation was divided into phases. First, soldiers would raid the district and apprehend high-value targets. Second, more troops would push into the district to help set up the concrete barriers, and finally they were to clear it out.

As the final planning developed, Peeples explained the mission to his team. He and Stout had a security element attached to them, and when the others started the raids, they would all move into position to cover the soldiers. Once the soldiers withdrew, Peeples's group was to be left behind, and from there they would snipe.

He did not have to explain the risk. Though given enough ammo, they would be alone in insurgent-infested territory most of the time. Hundreds of fighters were suspected to be there, and as hard-core as they were, they would probably fight to the death. Peeples knew his team would see some action, but he truly was not aware of just how much that would be.

To start the operation, the soldiers moved into the city under rain and darkness. They sped through the streets of Ramadi, knowing that enemy lookouts could compromise them. Peeples and Stout traveled in the rear Bradley, while the main element took the front vehicles. Before they hit their first target, the vehicles stopped just long enough for Peeples, Stout, and their security to exit.

Peeples kept his rifle in hand and rushed out. In the street, he noticed right away the double-stranded concertina wire lining both sides of the road. Stout pushed the wire down while the rest of the team crossed. A thought popped into Peeples's mind: "If insurgents want to sneak up on us, they will have to cross the wire as well."

Quickly the team infiltrated a building on the north side of the street. All of the action was to be on the south side and a good standoff distance away. In the hide, Peeples tasked the security as he always did; the Iraqi soldiers held the bottom floor to stay out of the way, while the U.S. soldiers took the upper levels. Within hours, the initial raid happened and the soldiers in the Bradleys exited the area, leaving the snipers to scour the darkness.

Behind his M24, Peeples searched out targets. The main road ran below his window, giving him a sight of an intersection 293 meters (just under a thousand feet) away. He'd been using night vision and had a great view, but nothing happened for the first few hours.

Later, a pathfinder element clearing IEDs swept through the area. Peeples covered their movement and heard gunshots intended for their vehicles. He spotted insurgents shooting from a high-rise in the distance, but they were out of his range. Peeples relayed the info to the pathfinders and the company. The pathfinders were armored and had not realized they were under fire. Once they learned of it, they moved out of the area. Thirty minutes later, four Bradleys arrived and took the direction of Peeples. The vehicles unloaded on the building where the gunmen were, and two of the floors collapsed on each other. The fighters did not return.

The rest of the night was quiet. The next day, Peeples, while action hungry, stayed on the gun. The others slept and occasionally scanned the area with him. The streets were empty besides a few cars. Their sector was dead and Peeples figured it to be an uneventful day—that is until around 1600, when a loud shaking of metal sounded nearby.

Everyone in the team heard it, but no one could distinguish the noise. It sounded to be nearby, and Peeples decided to look from the roof. The door leading to the roof was full of bullet holes, and Peeples peered through one to get a view. Outside, eight insurgents were trying to breach the concertina wire across the street. Two of them tried moving the razor wire while one held an IED. Another took the cord attached to the IED and went into a building.

Peeples ran down and called his team. He explained the noise and they instantly formed a plan. Peeples would initiate with his bolt-action rifle. Everyone else would take a predesignated target. The team also radioed the company and requested the help of the Quick Reaction Force; they asked the QRF commander to let them know when they exited the base, giving the sniper team about two minutes before they arrived. When the QRF commander radioed them, Peeples took aim.

The men with the IED were ninety meters away. Peeples chose the man orchestrating the event and put the crosshairs on his chest; at that distance he could have taken his head if he wanted to. The boom of his rifle started the attack, followed by the M240 machine gun and an M203 grenade launcher. The enemy was completely surprised and stood in shock for a second. With his first target down, Peeples transitioned to a guy trying to get the IED, which he had dropped in the street. The machine gunner kept on the few that slipped into buildings, until a minute later when the QRF arrived. As soon as they did, a counterattack was launched by more insurgents.

Gunfire ensued. Insurgents sent bullets through Peeples's window, causing him and Stout to dive for cover. During a lull, he stood and shot a man on an adjacent rooftop, but another volley flew through the window, hitting a metal windowpane and sending bullet fragments into his face and leg. The guy next to him took fragments to the face and hand. The impact shocked Peeples. It felt like a fist to the head, but when he knew that the injury was not serious, and neither were the injuries to the other soldier, they resumed fighting.

Five minutes later the shooting ceased. Peeples assessed the damage. Dead insurgents lay in the street and in the other building. One even dangled by his leg from a second-story balcony. Miraculously no soldiers were seriously injured, even though the wall behind them held fifteen to twenty bullet holes. When the coast was clear, the sniper team headed for the Bradleys to withdraw.

In the street, the team shuffled into the fighting vehicles. One of the soldiers tripped and was stuck in the concertina wire, sending Peeples to his aid. As he grabbed the wire, Peeples turned and noticed that they were being targeted by an insurgent. Before Peeples could level his rifle to shoot, a Bradley noticed the fighter as well and tore his body to shreds with its 25mm main gun.

At base, Peeples and his team resupplied. After a quick bite, they were ordered back into the fight. This time Peeples's security would be two squads, and instead of insert by vehicles, they would be trailing the company on foot. That night the soldiers patrolled from their base into their sector. Fighting had already begun in other areas, sending tracers and bullets twenty-five feet overhead while the soldiers moved in.

When they reached their break-off point, Peeples's team and the squads moved into a building. They held the upper floors of the six-story building with a commanding view of the city. From there, they watched as the rest of the company patrolled on and eventually out of sight. That night, the company took heavy contact while fighting to and from base. The next day Peeples and the others learned that one of their friends had died as a result. It was hard not to be dispirited. Another friend had lost his life, and Peeples thought about it all day.

The next evening, Peeples and his partner took the roof. An infrared beacon marked their friendly position for the air assets above that had made gun runs throughout the day. From the roof, Peeples kept his eyes in one sector while the others watched elsewhere. He and his partner scanned a certain road at the same time, and they noticed a large patrol moving in their direction. They focused in and saw two groups, the closest being 150 yards and moving in two single files, while the larger group patrolled about 500 yards behind them. When the snipers tried to recognize which unit the patrols were, they realized that these men were not friendlies; they were insurgents carrying AKs and dressed in street clothes.

"This has got to be friendly. There's no way this is enemy," thought Peeples.

Their size shocked him. He could not believe his eyes and even radioed the company three times to ask if any friendly patrols were in their sector. The answer each time was no.

Peeples immediately informed the company of the situation. He requested air and indirect fire but was denied because of collateral damage. With no help there, he turned to the QRF Bradleys and asked them to move toward his building from east to west, while Peeples and the squads covered the north-south street, locking the insurgents in an L-shaped ambush. He also requested illumination mortars to light up the sky on his mark.

Within minutes the Bradleys were en route. When they fired the first shots, the illumination mortars were sent and lit up the insurgents' position, causing everyone to open fire. Peeples ranged his first target at eight hundred yards. His custom M4 was capable, and he was glad that it was modified. With the light, he saw perfectly a man dragging a weapons crate. He adjusted his sights and let the first round fly. It hit its mark, dropping the guy, and Peeples swiftly acquired another target. His second shot was also on, and his second target fell, but soon the group dispersed.

Suddenly the lights went out for a few seconds. When the next mortar triggered, the illumination exposed an insurgent crawling. Peeples shot but was off target. He adjusted at once, but this time his bullet hit a power line. After he adjusted again, his bullet found its mark, stopping the man in place. Peeples finished him off with three more shots. Seconds later, he spotted another man resting behind a dirt mound. His upper torso was exposed and he looked to be taking cover.

"You see that guy behind the mound?" Peeples said, directing Stout onto the target.

"Got 'im," replied Stout.

Peeples's next shot may have been his luckiest ever. He aimed at the man's head and took a deep breath; he wasn't going to miss with the other soldiers watching. Relaxed and on target, Peeples gently squeezed the trigger. Stout and a few others watched Peeples's round hit the man's head, dropping him instantly. The sound of bullet hitting skull echoed through the street. It was a sound Peeples would never forget. Later the team pulled out and refitted back at base.

At base, everyone cleaned weapons and tried sleeping. They waited all day until night to insert into the city once more. This time Peeples and Stout were with a squad and were directed to hold a position in support of the soldiers erecting a concrete barrier. As luck would have it, Peeples and the squad were to be trapped in the Malaab District side while everyone else was on the other side.

The ride into the city was different this time. Insurgents in the Malaab felt the pressure and attacked the Bradleys the entire drive in. Once they were close to their objective, Peeples monitored the radio for the attached squad leader to signal that their building was clear. When it was, the Bradleys dropped their back ramps.

With his feet on the ground, Peeples knew right away it was going to be a tough mission. It sounded like the Fourth of July with the amount of explosions outside. All of the Bradleys were on the defensive as insurgents rained RPGs and bullets on them. Tracers zipped past Peeples and pinged off the armored vehicles. In the midst of this, Peeples and Stout ran for their lives and made it to the safety of the building.

The squad leader met them at the door. A situation report was quickly discussed, and Peeples took Stout straight onto the roof. Another soldier followed the two up, and once there, Peeples moved to the ledge facing the enemy and readied his weapon. His night vision covered his eye and he glanced over the ledge toward the street. To his surprise, below him were two insurgents with AKs, whispering to each other. Peeples's PEQ-2 infrared laser could only be seen with night vision, and he put the laser on their heads and dumped a half a mag, killing them. At the same time, another soldier fired on a few insurgents farther down the street. It was the start of the most violent firefight that Peeples had ever experienced.

Insurgents had staged in the same area that the soldiers

had moved into. Hundreds of enemy fighters opened fire very close to the soldiers. There was no room for mistakes now as the insurgents gave all they had. Peeples felt their bullets and RPGs exploding against his building, and even worse, he and his team were near the largest group, and the insurgents knew that.

The intensity of it all was nerve-wracking. It took all he had for Peeples to shoot, reload, duck, and shoot while trying to direct others on targets. His strategy amid the turmoil was to shoot muzzle flashes, but the enemy did the same and fired on him. Next to Peeples, the M203 gunner let grenades fly like a champion. He aimed for dead space where insurgents took cover and once landed a grenade behind a courtyard wall hitting several men. Their painful screams let him know that his shot was on.

It was a deadly stalemate for some time until the soldiers received air support. A Guided Missile and Large Rocket, or GLMR, momentarily silenced the fighting when one destroyed a house. Peeples gathered his thoughts in the mess of it all and hoped to make it out alive. As he regained focus, he saw an insurgent below him moving through an alley twenty yards away, and he killed the enemy fighter.

"These guys are close," he thought and threw a grenade into the alley. More insurgents appeared below them, and the soldiers tried pushing them back with more grenades.

"Allah Akbar!" yelled insurgents from a building across

the street. They also threw grenades back at the soldiers and peppered them with AKs. They began calling reinforcements, hoping to close with the soldiers.

The soldiers yelled back, unintimidated. They pinpointed the main building from which the insurgents regrouped. It was 75 meters (250 feet) away. With air on station, they dropped a five-hundred-pound bomb onto the building. The concussion rocked the neighborhood. Peeples went deaf for a second as debris rained down on his rooftop. When the dust cleared, a few insurgents crawled from the rubble still yelling, but they were easy targets.

The bomb gave the soldiers time to regroup. They were almost out of ammo, nobody had grenades, the SAW had no bullets, everyone was down to his last magazine of 5.56mm, and the machine gun had only a hundred rounds left. Fortunately for them, the soldiers in the rear assembled resupply pallets with ready loaded magazines, grenades, AT-4s (a light, anti-tank rocket), and machine gun ammo. Bradleys carried the crates to their building and arrived just before the fighting resumed.

The insurgents also regrouped. They pushed toward the soldiers in all directions, which transformed the machine gunner into a savage. He fired so much that his gun barrel turned red. The insurgents took a nearby building and positioned their machine guns from it, giving other fighters covering fire. The squad leader, realizing the situation, tore into the building with an AT-4, but though it was hit, gunfire still came from the house. Peeples grabbed another AT-4 and was so excited that he forgot to put the shoulder strap on correctly. When he fired, the launcher hit him in the face. Thankfully, in the darkness nobody had seen it. Unbelievably, the shooting did not cease, but a third and final rocket from the squad leader did the trick.

Soon, the soldiers noticed something peculiar. Every so often a loud shot, undeniably a sniper rifle, fired on them. The enemy sniper caught the attention of Peeples, and he made it his personal mission to find him. Peeples took his sniper rifle, used his PEQ-2 infrared laser, and flooded the windows he suspected the sniper was firing from. He had not heard the shot in some time, when out of the corner of his eye, he noticed a muzzle flash.

He scanned that area. Less than a hundred meters away, Peeples lit up an empty building with his beam. He saw a small circular glimmer of light, which was a reflection off the enemy sniper's scope. Peeples focused in and found the sniper sitting behind a table with his rifle resting upon it. The enemy sniper was clueless that he had been compromised. Peeples quickly aimed at his face and instantly killed the other sniper. It was the kind of shot that every sniper dreams of.

Soon the fighting spiked once more. The insurgents would not quit, but neither did the soldiers. Though he could not see him, Peeples heard Stout directing the machine gunner onto targets, followed by the sound of bullets penetrating flesh. He was glad to be next to these soldiers. They all held under fire with no sign of letting up. The fight started at 1900 and trailed off around 0300 the next morning. By sunrise half of the barrier wall had been erected and Peeples and Stout were taken back to base to debrief.

"You and Stout have been out for seventy-two hours. I know you want to go, but I'm keeping you here," said the company commander. He was proud of the snipers' actions, but he knew that they needed the rest.

Peeples and Stout looked like hell. They had not slept a wink the entire time and eaten only a small bit, and they had blood, sweat, and dirt caked all over. That night, after a hot shower and chow, Peeples fell fast asleep. The next day the wall was fully erected and Operation Murfreesboro was considered a success. Peeples was glad to have helped, but he was even more happy that the mission was over. He could not believe that with all the fighting, he had made it through intact.

Two weeks later, Peeples took R&R. He flew out of country and out of the war zone. It was just what he needed after all the action that he had seen. At first, he had to adjust to not having a weapon or being on guard the whole time. Though it was a break, he could not shake the thought of going back. He still had six more months to go, and if they were anything like the first six, he would have a lot more killing to do.

After a month, Peeples flew back. Instead of the war zone

that he had left with death at every turn, he arrived to the Anbar Awakening. In the time he had been gone, the tribes of the Al-Anbar Province had finally become fed up with al-Qaeda in Iraq and their devious tactics. Al-Qaeda's horrendous disregard for innocent lives pushed tribal leaders to side with the Coalition and defend their territory, and the result was staggering.

In his next six months the fighting was minimal. Peeples heard of only a few IEDs and no sorts of attacks of any kind. As a result, his team did not run any more sniper missions. It had gone from hell to heaven in one month.

In the end, Peeples learned much about sniping from his two combat tours. The biggest lesson was tactical patience. By using their heads, along with patience and good training, Peeples's team had been tremendously successful.

NINE

DUAL DEPLOYMENTS II

U.S. Army sniper Stan C. appreciated being home after Afghanistan. He had survived a helicopter crash and the fighting of Operation Anaconda in 2002, and now, being stateside on R&R, he could make do on his foxhole promises and be more friendly and enjoy time with others. He spent time with his family and friends, but sadly, the few weeks of recreation quickly withered. When Stan reported back to his unit, they began another cycle of predeployment training. Whispers among the battalion hinted at a war in Iraq, and Stan's unit would be involved.

First, though, Stan went to the U.S. Army sniper school at Fort Benning, Georgia. The school allowed his platoon three slots, and Stan was included. He and the other soldiers desperately needed to pass after their platoon sergeant stated that if they failed, they had no place in the platoon.

Stan and the two others arrived on the first day and dodged a bullet. After they had checked in, the instructors informed them that they could not attend because there was no room. Stan thought that his days as a sniper were over, until the instructors called roll and found that three others soldiers had not shown up. Knowing that Stan and the others had just returned from Afghanistan, the instructors cut them some slack and let them in.

Going through sniper school after having been in combat might seem like nonsense to some, but Stan enjoyed it. The more knowledge he gained, the deadlier he would be. School flew by, and Stan already knew the curriculum. Other students asked Stan about his time in combat, and he explained that he had not used his rifle that much but hoped to change that during his next combat tour.

After eight weeks, the three soldiers returned to their unit as certified snipers. The buzz among the battalion was still about a war with Iraq. It was definite now, but the soldiers had to wait until they were directed to deploy. Until then, the snipers would shoot, pack, and inspect their inventory while bearing the "hurry up and wait" mentality. Finally, though, in early 2003, the unit was on the move.

The 101st Airborne Division began shipping equipment to the Middle East in February 2003. Blackhawk, Apache, Kiowa, and Chinook helicopters, along with land-based vehicles and equipment, were sent for Operation Iraqi Freedom. By March, the 101st, capable of deploying infantry air assault and housing two aviation brigades, was fully equipped and ready for war.

In Kuwait, Stan's unit was sent to Camp New Jersey. It sat twenty-one miles southeast of Iraq, surrounded by nothing but sand. Everyone settled in and waited for the war to start, and in the meantime, more training was advised. Stan's team ran contact drills mimicking firefights, to instill instant reactions. They patrolled the perimeter with heavy packs to gain leg strength and to fine-tune their weapons, shooting their sniper rifles at every chance.

Morale was high among the soldiers in the battalion. Under the V Corps, the U.S. Army's spearhead and main element for the attack, the soldiers knew that they were making history. First Afghanistan and now Iraq, and the talk of fighting brought comparisons between the two war zones. The soldiers had gone toe-to-toe with the Taliban and al-Qaeda, and they were tough. From everything that was said, it looked as though fighting the Iraqis would be nothing compared to the enemy in Afghanistan, but they would have to wait and see.

Stan was anxious to get started. His new spotter, Steve, was also a veteran of Afghanistan. They pushed each other to get better as snipers and talked over every possible combat scenario. The longer they waited, the more Stan thought about combat. This war would not be a cakewalk, and Stan felt a healthy mix of fear and excitement. He figured, though, that it could not be any tougher than his last deployment.

Finally, for the snipers, the training and boredom came to an end. One day, Stan sat talking with Steve in his tent. With painted weapons and stuffed packs, there was nothing left to do but wait for the call to move out. Suddenly, their platoon commander walked into the tent.

"OK, boys, this is it. Break down everything and get it packed up. Be ready, because in the next twelve hours we are crossing the border."

Stan shot Steve a look of excitement and they quickly snapped to the order. A familiar feeling of butterflies settled in Stan's stomach. He and the others positioned their necessary gear and left everything else in a pile near the center of the camp. Stan's team supported Alpha Company, and Stan met the company commander at their staging area. They discussed how the sniper team would be used, and it was determined that while the company moved, the snipers would be guardian angels providing over-watch. If needed, they could conduct conventional sniper missions as well.

That evening the men enjoyed a final night of good sleep. The next day, Stan and Steve were up early when suddenly sirens sounded in the camp. They indicated a missile attack. When he heard the noise, Stan reached for the gas mask strapped to his hip and had it on in ten seconds. Nearby, Patriot missile batteries launched to intercept the incoming Scuds.

The attack brought apprehension. The Taliban had not used Scuds. Stan began contemplating what they were up against, and a few hours later the order to cross the border was given.

Al Anbar

Marine sniper Josh Rush returned from Afghanistan ready for another shot at sniper school. He had failed the first time during land navigation, but navigating the mountains of Afghanistan had given him the experience needed. Shortly after returning to Kaneohe Bay, Hawaii, Rush checked in to the Third Marine Division sniper school once again. Combat operations taught him what he needed to pass sniper school, and ten weeks later, Rush was a certified sniper.

His platoon from Second Battalion, Third Marines dwindled after Afghanistan. Senior snipers discharged, leaving room for the junior marines to step up. Rush soon became a team leader responsible for three others. Evers assumed the role of platoon sergeant, chief scout/sniper, and First Team leader, but his acceptance to a position at sniper school left Rush to fulfill the duties. Rush had his hands full, especially because his battalion was deploying to Iraq.

Rush's platoon underwent predeployment training. They tailored their tactics to urban environments and for IED prevention, completely different from what they had faced in

Afghanistan. Sniper school and other reports taught the snipers that their major challenge in Iraq would be stopping IEDs. The sophisticated manner in which the insurgents employed the bombs called for a very basic and fundamental element to prevent them—an element that snipers learn from the beginning of their training: observation.

Before leaving, Rush learned of his new area of operations. His battalion would cover Haditha, Haqlaniyah, and Barwana, the three towns making up the Haditha Triad in the Al Anbar Province of western Iraq. When he learned that, Rush knew right then that it would be dangerous. Previous fighting between Coalition Forces and insurgents was well broadcasted there. More importantly, Rush knew that in the area, an entire six-man sniper team had been killed. The incident forced many sniper platoons to reexamine their employment and tactics. It was also a warning that snipers in Haditha needed to be ever wary.

A year earlier the Third Battalion, Twenty-fifth Marines were stationed in Haditha. Security and stabilization operations had brought the Ohio-based reserve unit to Iraq, and they were anxious to fight. Insurgent groups operating in the area quickly met the marines. The snipers passed the time running counter-IED, reconnaissance, observation, and surveillance missions. Teams were successful at stopping IEDs in certain areas, but as the end of their time in Iraq approached, their base was routinely harassed by rocket and mortar fire. After one occasion, sniper teams were sent on a mission to observe the suspected firing locations.

During the operation, something went terribly wrong. The six-man team was inserted by vehicle near their objective. Another team was sent to observe elsewhere, only three kilometers away. According to Ansar al-Sunnah, a terrorist group operating in the region, the mortar and rocket fire upon their base was bait for the snipers, drawing them into the area. Once there, the snipers were ambushed and killed. It is not known whether or not the marines were set up or exactly how the insurgents were able to get close to the snipers without drawing fire. The fact still remained that insurgents were aware of snipers and how to manipulate their operations.

For Rush, the event reinforced his need for stealth. He knew that in Iraq, the enemy was not to be taken lightly.

The Push

A massive convoy of vehicles flooded a makeshift road into Iraq. The movement caused fine, white, powdered sand to cover everyone and everything. It clung to Stan's entire body and gear. Unfazed, Stan was fixed on fighting and surviving. He had no clue what his time in Iraq would present.

After a day of traveling, the soldiers stopped west of Basra. No enemy contact had been made yet, but reports indicated that enemy armored units were in the area, possibly to stage an attack on the soldiers. The company commander unleashed

Stan's team and others for long-range reconnaissance. For a week, Stan carried his M24 sniper rifle beside him in the Humvee in which he scouted the area, but there was no sign of the enemy. His unit was tasked with an air assault into Hillah.

Al-Hillah was the central city of the Babil Province. At the time it supposedly hid hundreds of enemy fighters. Saddam's elite Republican Guard were the troops there, and Stan's unit entered the city.

The city was also empty of fighters. Stan climbed into a building and found a great position to cover the others patrolling. In the room, he found two shell casings the same size as the ones used by his sniper rifle. When he checked the bottom, he saw that they were American made. He wanted to use his own bullets, but hours passed before his unit moved on. While leaving, they were allowed to ditch the heavy chemical-protective suits. The relief was felt by everyone.

Shortly thereafter, the soldiers flew into Saddam International Airport. Stan's team arrived by helicopter and positioned themselves in the terminal to observe the city. The airport had already been secured by the Third Infantry Division, but firefights continued throughout Baghdad. Later, his unit entered deeper into the city, on their way to the Ministry of Finance building.

Charlie Company took the lead and Stan's sniper team accompanied them. Other U.S. military units waved as they passed, and civilians did as well. The idea of being downtown

Baghdad was exciting. Granted this was not a vacation, but Stan still appreciated being there.

The soldiers reached the eighteen-story building at night. Levels 14 through 16 were still on fire from looters and bombings. Regardless, the structure was made for sniping. Possible hide sites were unlimited. Countless windows were built into the two identical wings that formed the building.

Stan knew what to do. Inside, he moved positions on different levels throughout the day, and at night they moved to the roof. Up top, a chest-high wall allowed the snipers to stay concealed while they used night vision to observe. At night, staring at the thousands of city lights, Stan contemplated the differences between here and where he had been. In Afghanistan, he never entered the cities, only the mountains. There, often his team was alone, with nobody around for miles. In Baghdad, he was in a city with millions. The change was drastic.

At their new location, the soldiers had a fairly easy task. A nearby suburb housed members of Fedayeen Saddam, a group of volunteer citizen fighters loyal to Saddam Hussein. To catch them, rifle platoon squads held checkpoints on the roads while the snipers covered them. The snipers found that an overpass spanned horizontally between them and the neighborhood. Below that, another road led straight from the snipers' building into the neighborhood, connecting with multiple side roads and intersections.

In the building, the snipers created a system for operations.

By now, Stan and Steve were accompanied by two fellow snipers. One soldier was able to rest while the others traded spotting, monitoring the radio, and manning the sniper rifle. Below them, Charlie Company soldiers spread out to different locations, some holding the bottom floor.

Within the week, Stan made his first kill behind a sniper rifle. Papers and other materials littered their second-story hide. Stan's M24 lay on his pack and rested in his shoulder. In the middle of the day, a green car passed Stan's position. He noticed it, but he was focused on the neighborhood in the distance. Suddenly, from the car, shots rang out. A man leaned from the passenger window aiming at a Charlie Company checkpoint.

"My dope is at 300," Stan reminded himself. He was not about to let the men get away.

No time to waste by adjusting his scope. The car was speeding in the opposite direction, and Stan briefly noticed the passenger's rifle barrel dangling from the window. He estimated that the car was past three hundred meters and he placed the center of his crosshairs just high, above the passenger. The shot broke the silence in his building.

As habit, Stan chambered another round while the recoil ceased. He saw the back windshield shatter, and the barrel of the AK, which had been sticking slightly from the passenger window, slumped. The passenger jerked forward, and seconds later the car drifted to a halt.

Moments later, soldiers from the checkpoint reached the car. Over the radio, a soldier asked Stan if he had taken a shot. He confirmed. They verified a hit and that the passenger was dead. Watching his friend die in the next seat had terrified the driver so much that he had stopped driving and sat motionless.

Having stopped the shooter satisfied Stan. It also sent a message that U.S. snipers were always watching in that sector. Even to other soldiers the shot was impressive. Later that night, Stan descended to meet Charlie Company's first sergeant for resupply on the first floor and was met with praise.

"That was a hell of a job. How did you shoot a guy in a moving car?" asked the senior soldier.

"I don't know. Ask your brother, First Sergeant. He's the one who taught me how to snipe," replied Stan. The first sergeant's brother was an instructor at the SOTIC. Stan knew that he had no room to brag. To him, it was a lucky shot.

The next morning, the snipers continued their routine. Traffic passed by their building all morning, and in the afternoon, while Stan searched the neighborhood with the spotting scope, he noticed a van parked beside an orange and white painted taxi. Two groups of men loaded the vehicles. At first Stan didn't notice anything suspicious. As the men approached, however, he realized that all of the men inside the vehicles held weapons.

"I've got something here," Stan informed his team.

The snipers knew the routine. One soldier, Adam, quickly

mounted the Barrett .50-cal sniper. Steve reached for his M4 and Stan would remain on the glass, spotting for the two.

Moments later the vehicles stopped on the street below their building. All at once, the men inside them opened fire. Two men jumped from the car and started to hammer the soldiers in the bottom floor. With the gunfight reverberating throughout the area, Stan's teammates desperately wanted to help, but there was a problem. The snipers had no angle to engage. They were so high in the building that if they wanted to shoot, they needed to lean out of the window. That, however, would have given away their position.

Meanwhile, Charlie Company soldiers returned fire. When they did, the drivers turned their vehicles and began to drive back to the neighborhood.

"Don't let that taxi get away," Stan said to the others. He was fixed on the car because it was the faster of the two vehicles. The gunmen sped away, but they were approaching one of the snipers' predetermined target reference points, an intersection at 650 meters (a little more than 2,000 feet). With that in mind, Stan told Adam and Steve to adjust their sights and to stand by for a wind call.

Seconds later the car reached the intersection. The driver swerved to miss other cars and hit the median, stopping momentarily.

"Hold center," said Stan. It was their code, meaning that no adjustments for wind were needed. As he announced that, next to him Adam fired one shot with the .50-cal. The crisp, sunny day allowed Stan to see the vapor trail from the bullet. The bullet entered the front right panel, just above the tire, causing the car to shake lightly.

"Hit! Fire again!" said Stan.

Adam turned to the passenger. Stan's spotting scope magnified his view to 10 power. With it, he clearly distinguished the passenger's face. The man gazed in their direction to find where the shooting had originated. While he did, Adam sighted in on his head and fired once more. Stan witnessed the second bullet remove that man's brains from his skull, sending chunks of them everywhere. Simultaneously, Steve put in work as well. He shot the driver twice and the back passenger once.

Meanwhile, Charlie Company soldiers made quick work of the van and everyone inside. Other soldiers rushed to the scene and inspected both the car and the van. Stan watched through the scope as they pulled the dead from the vehicles. He teased Adam, saying he wanted credit for his kill as well. Of the three, Adam was the worst shooter, and without Stan spotting, he would have never made the shot. Adam was unconcerned, and a little sniper banter never hurt.

With two impressive performances, the snipers gained notoriety among the battalion. It gave the commander added confidence to let the snipers operate independently. That trust helped Stan's team for future missions, especially after leaving Baghdad.

Haditha Triad

The Second Battalion, Third Marines arrived in western Iraq in 2006. Haditha was their new sector. Their base was a hydro-electric plant built into the Haditha Dam and powered by Lake Qadisiyah. With the information of past attacks and that snipers had been operating in the region for years now, Rush's platoon understood how to adapt. Insurgents knew the streets and roads and likely positions that U.S. forces occupied. To stay undetected, the snipers would have to try new locations.

Fortunately for Rush, his battalion commander gave the snipers freedom to operate. He had learned that the snipers were better off supporting the battalion rather than the individual companies. It allowed them to move throughout the area without having to abide by company commanders' directions.

Right away, Rush learned that the area was hot. His platoon settled in and started changeover with the snipers from the existing unit. The team leaders accompanied the snipers on missions to give them a lay of the land. On that first mission, the snipers dropped an insurgent trying to plant a bomb near the road. It was the first of many killings for the snipers of 2/3.

Soon, a change of operations gave the snipers a chance to demonstrate their skills. Along Route Bronze, a main supply route leading from Ar Ramadi to beyond Haditha, armed checkpoints had been scattered up and down the road. Their purpose was to stop the massive number of IEDs that were being emplaced on it. When Rush's battalion took over, casualties had begun to mount, and the marines no longer had enough manpower to supplement the checkpoints. They were forced to abandon many of the checkpoints. It did not take long for insurgents to catch on, and within the week, Route Bronze was laden with IEDs.

On their first mission to stop the IEDs, Rush's team infiltrated the desert near Route Bronze. Having been in the city for some time, the snipers needed to adjust to their new environment. Flat, open desert lined Route Bronze on both sides for miles, making it hard to build or find a suitable hide. Camouflaged netting concealed the team, and the snipers dug into the hard ground.

The next day brought enemy activity. The snipers were able to see for miles in the open desert. Rush witnessed a man digging on the road, but he was 1800 meters away (just under 2,000 yards). Rush only had the M40A3, with a maximum effective range of 1,000 yards. The .50-caliber sniper rifle would have been more useful, but they had not brought it. Rush decided to engage anyway, hoping to get lucky, but the distance was just too great and the man escaped with his life. Rush would have to wait until his next mission to kill bad guys.

After a few weeks there, Rush hated the desert. The freezing temperatures made it miserable. His second mission called for his team and another team to mutually support each other while keeping eyes on a certain area. After insertion, the snipers found a great hiding position and decided to co-locate there. They positioned themselves under mounds of sandbags and pieces of barriers from one of the checkpoints. It took them some time to dig in, but by morning, all of the men were completely hidden.

While other marines slept and held security, Rush and another sniper took watch. In the afternoon, movement on the road drew the snipers' attention. Rush focused in with his spotting scope and watched as a station wagon stopped. One man got out and began pouring gasoline on the road. This fell under insurgents' tactics, techniques, and procedures for planting bombs. The gasoline softened the asphalt enough for it to be dug into, allowing IEDs to be placed underneath.

When he saw this, Rush reached for his laser range finder. Disappointment and anger were his reaction when he saw that the vehicle was 2,200 meters (7,200 feet) away. He watched the man for a minute while the radio operator sent back a report of the activity. The snipers grew frustrated, knowing that either the man would get away, or the patrol would catch him and arrest him. Considering that he was trying to kill Americans, and as many as possible, the two scenarios did not seem like justice. In a flash, though, a new scenario appeared.

Rush's eyes lit up when he watched the man get back into his car and drive toward the snipers.

"Get up, boys. Grab the SAWs and move over here," explained Rush.

With seconds, two snipers sat up, ready for the station wagon to get within range. On Rush's cue, the marines opened fire. It took less than twenty seconds for both SAWs to kill the driver and mangle the car. The incident stopped IEDs in that particular sector for a short while.

Within the platoon, other sniper teams were successful as well. One sniper team in another area stopped five men from planting IEDs. They took to a house near an intersection, and one day a car stopped and men jumped out to try planting bombs. The snipers opened fire and killed four of them. One man escaped, but a blood trail was left.

As their time wore on, the marines were able to stop more IEDs. Airborne communications jammers flew certain routes daily, sending electronic signals along the way. These signals became so good that insurgents reverted back to laying wire to manually detonate bombs.

The roads around Haditha were ripe for insurgents to plant IEDs. Open desert allowed them to see military vehicles approaching. There were also roads that had dead space near them, giving IED cells access to move onto the roads and to slip away. Wadis were a great way for them to move near roads, as well, and once Rush stopped a man near one.

His team had been in place for two days. Their hide was a small hole near a road. Rush's teammate was on watch when he spotted a man. The person they observed pulled wire from one position to a nearby bridge. It was blatantly obvious that he was going to use it for an IED.

"What's the range?" asked Rush.

"Six hundred," replied his spotter.

Rush dialed it onto his scope. When he aimed in, he saw that the man was walking down a draw. Rush lined him up and fired one round. His spotter called a miss, just off the man's left shoulder. The man had no clue as to what was happening. He heard the shooting and stood still.

"Hold right shoulder," said Rush's spotter.

Rush held to the right and fired again. His round impacted center chest and killed the man where he stood.

A few months later, Rush felt the effects of the Anbar Awakening. Local tribes and leaders began to resent al-Qaeda in Iraq and other terrorist groups. They'd had enough of indiscriminate bombings and killings by the terrorists and joined with the Coalition to stop them. Some of the men who had attacked marines before now sided with them. For Rush, the awakening saw the end of his combat operations.

Northwestern Front

In 2003, by summer's end, the Second Battalion, 187th Infantry were relocated. They were shuttled by helicopter from

Baghdad to a small town on the northwestern front known as Sinjar. Open desert surrounded this remote community with less than one hundred thousand people, and the Sinjar Mountains could be found immediately north. A single road ventured from Syria into the town and ran farther east, into Mosul.

The battalion set up headquarters west of town. Their mission involved preventing the influx of foreign fighters passing through Sinjar. There, Stan's sniper team started missions almost immediately.

Informants indicated that a mobile black market weapons exchange took place in the town. The dealings happened from the trunks of cars and trucks. Stan's team sat in on the intelligence brief and was directed to get eyes on the event and report all activity. Planning was easy. After studying the terrain, Stan decided to emplace his team within a grain silo near the market. His M24 would be in range from there. On the night of the operation, though, as they patrolled into the area, Stan noticed a cemetery and decided to set up there instead. They hid among the clustered gravestones, with a clear perspective of the marketplace to gather intelligence for the battalion.

They reported several incidents on their first day of observing. Twenty vehicles arrived, many loaded with weapons, ranging from AKs, to pistols, ammunition, and other small arms, but no crew-served weapons or RPGs. Stan also reported that near the trading site, several men carried weapons to and from houses. The battalion wanted more evidence, specifically for heavier weaponry.

By the third day, the snipers saw exactly what the battalion wanted them to see. Nearly seventy vehicles crowded the square. Russian-made PKM machine guns, RPGs, and AKs were for sale. When the snipers reported the situation, the battalion initiated their plan to raid the marketplace.

Twenty minutes after the report, Apaches led the way. A company of soldiers followed minutes behind. Confiscating the weapons and apprehending the people was their concern. The snipers recognized the buzz of approaching helicopters, but so did the men in the market. They loaded their vehicles and drove east in a massive convoy.

Stan could not engage the men. The plan did not call for target reduction. Instead he warned the raid force that the men were escaping. Apaches, the first on the scene, met the convoy and intervened, preventing the vehicles from going farther. Soon, the rifle company arrived. With the direction of the snipers, they raided houses. By nightfall the soldiers had recovered mortars, RPGs, machine guns, explosives, ammunition, and small arms.

As a result, the soldiers realized that they needed a foothold in town. One company was sent to assume the town's old police station. The battalion planned to always keep one company there and rotate others in and out. When the town's insurgents learned of this, they threatened to attack the soldiers and warned them not to move in. The soldiers never backed down. Once they learned of the insurgents' plans, they sent more soldiers in, to include the entire sniper section.

Stan arrived at the police station along with the other snipers. He grabbed his weapons and positioned his team on the roof. The other snipers were there as well, but each team held a separate sector. Below, soldiers patrolled the streets and neighborhoods. Everyone was to be on alert all night to find out if the insurgents would come through on their promise. And they did.

From Stan's position, he covered directly across the street to a strip of shops and alleys. The town's lights lit up the quiet streets, so much that the snipers had no need for night optics. Nearby, only 100 meters away (330 feet), a mosque with a minaret was covered by another sniper team. Stan and his partner traded watch through the night. By 0200, they expected that the insurgents were not coming. Suddenly they attacked in force.

RPGs shattered the silence and hit the Humvees below. AKs and machines guns immediately followed. The rockets came from Stan's sector, and he instinctively reached for his M4 while his partner did the same. They searched the area in front of them and found two men in an alley. No words had to be said, and the snipers aimed in on them.

Elsewhere, soldiers and insurgents traded fire. Stan tracked one of the men in the alley less than a hundred meters away.

The man moved into the light, revealing his entire body, giving Stan enough time to rest his sights on the man's chest and fire. His bullet entered the man's chest and sent him onto his knees. Stan quickly fired again, dropping the man onto his back. Next to Stan, his partner aimed at the other man, who attempted to drag Stan's target away. He was killed with one shot.

From the minaret, a machine gun fired on the soldiers. Another sniper took aim at the machine gun's muzzle flash and shot once, silencing the weapon. Stan and his partner searched for more targets, but they were out of range. Soldiers on foot patrol were fighting nearby, and Stan searched in their direction.

Ten minutes later, a man stumbled from the mosque and into the road, holding his stomach. It was the gunman from the minaret and he was hit. A squad close by searched the man as he lay dying; a single bullet had entered just below his sternum.

The fighting lasted fifteen minutes. Later, the soldiers raided the town's hospital to interrogate the wounded. There, soldiers found the men whom Stan and his partner had shot. One died while the other was in the process of dying. The man who had been shot in the sternum wore a blue Iraqi Police uniform. The soldiers realized that none of the police could be trusted.

As Stan spent more time in Sinjar, his team assisted with many different missions. When they were needed for raids or reconnaissance, Stan was allowed to ride in helicopters to scout objectives from the air. He rode with Blackhawk helicopters to scope the lay of the land and get a view of the best sniping positions.

One day Stan's team was ordered to observe and report on a certain target in a nearby village. Their target was a village elder with ties to the insurgency. The battalion learned that the man's son had died in a shoot and his body had been transported back to the village for burial. Stan's team needed to infiltrate the village and report on the man's movements before a raid force was sent to apprehend him on the morning of the son's burial.

Stan's team patrolled in on foot. With absolutely no ambient light, the pitch darkness made it hard for the snipers to find a suitable position. They crept between trees and shrubs two hundred meters from the target's house, but nothing would hide the four of them. Stan had an idea. The cemetery was built into a nearby hill and Stan led his team there.

In the graveyard, Stan decided his team would stay there. They stumbled upon a freshly dug grave big enough for all of them to occupy. It was unorthodox for the snipers to set up in such a common area, but the position had a suitable egress route and the snipers were low enough that it would be hard to see them, especially that night. They also knew that they were not going to be there long. Early the next morning, a raid force would move in while the village gathered for the funeral.

Stan reached for the PAS-13 thermal night sight in his pack. With the night vision rendered useless, the PAS-13, able to detect heat, was used to observe. The village had no street-lights, and hours passed with no movement. Stan felt that something was amiss. Normally some movement was observed, and Stan felt that their mission might have been compromised. He reported that the raid force should move early. Within the hour, soldiers arrived in helicopters, and the mission went as planned. During debrief, Stan learned that his team had set up in the grave in which the elder's son was to be buried.

In November of 2003, Stan's team was sent out on another adventure. Third Armored Cavalry Regiment (ACR) patrolled the Al Anbar Province, south of Stan's battalion. The ACR requested a company to support their push into the border town of Husaybah. Charlie Company was elected, and the commander wanted two snipers attached. Since their area was quiet, all of the snipers wanted to go, but after a brief discussion the team leaders decided on Stan and his partner. The two of them packed up and met with Charlie Company to convoy south.

After a long drive, the soldiers met with the Third ACR. The plan was to patrol through the small towns on their way to Husaybah. The snipers were to be used as over-watch, and when they finally made it to Husaybah, they would sneak into the city in advance of the main force. To Stan, it was nothing new. He had done plenty of it before.

Two weeks into the operation, Stan and his partner were alone in Husaybah. They were blocks ahead of the main force, who were clearing houses in Stan's direction. He and his partner occupied the top level of a gutted and partially built three-story house. Inside, Stan and his partner held security for the advancing force and had been doing so for hours.

Outside, the sky was gray and dark. It had rained for nearly a week.

"Hey, I'm going to the bathroom," Stan said to his partner, and he reached for his weapon. Their headsets allowed them to keep communications.

Stan slipped into another room, careful not to draw attention to himself. There, he noticed movement outside. With his scope, Stan caught sight of two men moving through an alley in the distance. Nobody else was on the streets. He told his partner that he was going to investigate the men. In another room, he gazed through a window and noticed what the men carried. One had a PKM machine gun, while the other carried an AK with a bundle of RPGs strapped to his back. When Stan saw them, they were running from left to right, but they suddenly turned and headed straight toward his building.

"We're compromised," Stan said, worried. He remembered the Special Forces snipers that had fought and died in Somalia after being surrounded. He believed that his fate would be the same.

Without hesitation, Stan leveled his weapon. The men were

easy targets. They ran straight at him. After Stan shot both men in the chest, his partner asked him the situation. Stan met him in the room, telling him to pack his gear. When they were ready, the two snipers moved out of the building and linked up with the other soldiers.

When the operation was complete, Stan moved back with his battalion to Sinjar. He spent the next six months conducting sniper missions and eventually his time came to an end. When it was all said and done, Stan felt lucky that no one in his section ever died. Others in the battalion had succumbed to insurgents in Iraq and the fighting in Afghanistan, but no snipers from his platoon.

TEN

ROE

Rules Of Engagement—Directives issued by competent military authority that delineate the circumstances and limitations under which United States forces will initiate and/or continue combat engagement with other forces encountered. Also called ROE.

—**Department of Defense Dictionary of Military and Associated Terms**

NEGOTIATING the streets of Iraq is tough. Amid fighting the insurgency, evading IEDs, repressing ambushes, and surviving attacks, snipers face another obstacle, one that can be as confusing and complicated as the fighting itself. It is the Rules of Engagement.

The purposes of the ROE seem simple. They regulate the actions troops can and cannot take in hostile situations, but for snipers they can also become hindrances for a few reasons.

First, snipers are often used for prevention, which calls for split-second, life-or-death decisions to decipher intent. These are tremendously hard decisions to make under some circumstances. Secondly, insurgents know exactly what the U.S. military ROE are and have tailored their operations to skirt them in order to accomplish their missions. Lastly, because of the ROE, many snipers fear making any decision, knowing that if their actions seem suspicious, the repercussions can destroy their careers. One Marine sniper in particular, Sergeant Johnny Winnick II, learned this lesson the hard way.

In 2007, Johnny was aboard the USS *Bonhomme Richard* as a scout/sniper team leader with the Third Battalion, First Marine Regiment. At the time, "The Thundering Third" made up the ground combat element as the battalion landing team for the Thirteenth Marine Expeditionary Unit. He and the other marines had been out to sea for months and were in much need of ground time. Fortunately, by May, their Western Pacific float steered them toward the Middle East and to a four-month stay in Iraq.

Before debarking, the marines sat in on the typical briefs. They were informed of their mission; it was ground combat counterinsurgency operations in the Al Anbar Province. The insurgency was alive and well there, and Johnny knew it, having been there twice before. His team sat quietly listening to the intelligence officer explain cultures and customs, what they could and could not do. Finally, he described the Rules of Engagement.

Johnny's team knew the rules of combat. Most of them had been to Iraq at least once. This was Johnny's fourth deployment there; twice before he'd gone as a machine gunner and once as a SAW gunner in a sniper team. Stabler, from Missouri, was the assistant team leader and this was his second deployment there. The two of them had passed sniper school together just months earlier. Alex, the team's radio operator, had been to Iraq, as well. They knew what to expect of the brief, except this time, when the officer was finished the marines were struck by one sentence.

"Remember this, gents, the Marine Corps eats its young," said the officer.

He was referring to the fact that if the team screwed up in Iraq, they would be prosecuted to the fullest extent. They all knew what he meant, after having witnessed what had happened during their last deployment, when marines from their battalion were accused of murder in Haditha, Iraq. None of them wanted to be on the receiving end of that hell storm.

A month later, Johnny and his sniper team were punched out to FOB Golden, near Lake Tharthar in the Al Anbar Province. They were quickly spun up on the area and the enemy situation. Insurgents owned the area, particularly al-Qaeda in Iraq, whose members roamed freely, establishing training facilities and terrorizing U.S. collaborators. Though small arms and mortar attacks were regular, the biggest danger facing the team would be, by far, the immense amount of IEDs.

The area was a haven for IED factories and bombings. Small teams of insurgents made up IED cells whose sole purpose was to make, plant, and detonate the deadly bombs. The most common were artillery or mortar rounds planted near or on roads. Even more deadly were suicide bombers in vehicles packed full of explosives, attempting to get as close to the marines as possible before blowing themselves up.

Other threats came in daisy-chained IEDs, where more than one bomb was linked together for multiple explosions. Some were pressure plated and positioned in the road, destroying vehicles when they were run over with the slightest of weight. The snipers also learned of a new, very powerful bomb made of ammonium nitrate packed in jugs, barrels, and cars. The jugs and barrels were a good way to hide the contents from snipers, who, the insurgents knew, needed positive identification before shooting.

The snipers were also told of the insurgents' increasingly sophisticated tactics. They had planting IEDs down to a science, literally emplacing them within sixty seconds. Unmanned aerial vehicles caught footage of their method, which was broken into three parts. The "softeners" arrived first with gasoline or other substances to break up the concrete, asphalt, or dirt. Next, the "diggers" hit the exact same spot, making a hole for the final component, the IED "layers." Ideally, this method endangered only the IED planters, allowing the softeners and diggers to escape because they were unarmed. The snipers realized the challenge ahead of them, but relished the idea of stopping IED layers and saving others from being hurt by them.

Johnny's team was anxious to get started. Their first mission was simple; they were ordered to provide surveillance on a mosque near their base. It was suspected that IED cells and bomb-making materials came from it. Should they spot anyone taking IED materials from the building or laying bombs in the road, they were to get permission and engage.

The mosque was not far away, only fifteen hundred meters (five thousand feet), close enough for the team to hump into position, which they did the next night. While departing, Alex radioed the COC, or Combat Operations Center, for a radio check. They were only a few hundred meters out, but Alex knew something was wrong, because he could barely hear the marine on the other end. It was a consistent problem that would plague their entire stay.

Early the next morning, Johnny's team was in place. The radio had given them trouble the entire time, but they kept on with the mission. Hidden, the team was observing the mosque from a nearby building when, in the afternoon, someone spotted suspicious activity.

"We've got company," said the sniper on watch.

A mix of cars and small trucks stopped on the road next to

the mosque. The drivers spilled out and quickly gathered around the back of one car. Johnny took over the gun and followed in his scope, noticing that a few men pulled IED materials from the trunk and placed them in the ground.

"Get the COC up, and let 'em know what's going on!" he shouted, needing permission to engage.

Alex tried to raise the COC, but the radio would not have it. He immediately switched to the iridium satellite phone, their emergency communications device, but the Iraqis were gone before he could explain everything.

Frustrated, Johnny directed a nearby quick reaction force to the IED's location. Sure enough, the marines found a 155mm artillery round fashioned into a bomb. To make the situation worse, when the snipers arrived back at the base, Johnny was lectured for not taking the initiative and eliminating the IED cell. He was told not to hesitate if the situation happened again. Being reprimanded for something that he could have prevented was disheartening, especially because he was obligated to get permission before engaging.

All that aside, if the exact situation arose again, he knew what to do. He had seen too many casualties from IEDs, and if he had the power to stop it, he was not going to be the sniper who allowed people to get injured or killed by any more.

Days later, Johnny and his team were assigned to the same mission. This time they planned to observe from a different location. They slipped off base early on June 17, under the cover of darkness. The patrol lasted longer than expected because of an abnormal number of convoys on the roads. Humvees mounted with lights shining in 360 degrees were the ones to avoid. The snipers knew that the gunners inside shot at anything they saw, and everyone hit the ground when they passed. Alex kept a pop-up flare handy, a signal meaning "friendly," just in case they took rounds. He was already annoyed that the radio was not working again, but he had not had enough time to give radio checks before losing contact.

Shortly before sunrise, Johnny and Stabler stopped the team to recon possible hide sites. The satellite imagery they used to mark an initial position deceived them. The designated building was actually a small platform only a few feet off the ground and was not big enough for the entire team. With little time, Johnny and Stabler searched a nearby structure. Inside, the feel fit the description of buildings used by terrorists to videotape their killings. Bullet holes sprinkled the walls, along with Arabic writings. There were also plastic sheets on the ground.

"Is this a kill house?" Stabler asked.

"It looks like it," Johnny replied.

Though they did not want to stay there, the sun was about to rise and they had no other options.

Johnny called the team in and everyone prepared the hide. Immediately there was a problem. From the building, they could not see the all of the primary objective, the mosque.

Several buildings were in the way, making it only half-visible. The gas station, a meeting point for insurgents and a known IED hotbed, was the secondary objective, and it was in plain sight. The secondary objective would now have to become the primary.

In the hide, communications took priority and Alex went to work. He sat against the wall and pulled out the radio. Johnny and the others set up security, except Stabler, who prepared the shooting position. The front doorway had the best view, and he placed the rifle there facing the gas station. He found that the area could not fit more than one person, so whoever manned the gun would have to relay everything to the team.

The snipers were in place and ready to observe by sunup. Stabler rested Johnny's MK-11 on a tripod and took first watch. He had screwed on the suppressor for Johnny, who preferred the rifle over the M40A3 because of its ability to put rounds downrange faster. Stabler did not care what weapon he had just as long as he could shoot it. Johnny sat a few feet away with his Mossberg 500 pump-action shotgun handy. The MK-11 allowed him to carry a close-range weapon, and he preferred a shotgun. Josh had an M4 with an M203 grenade launcher, while Alex carried an M4.

Shortly after sunrise, the action picked up. The gas station was three hundred yards away, next to a four-way intersection with a stoplight that maintained traffic in all directions. Stabler scanned the area and noticed a white Toyota pickup with six men stopping on the side of the road. The front passenger got out, walked to the back, and opened the tailgate. Four men in the back looked nervous while the front passenger grabbed something and walked a few meters behind the truck.

"Hey, boys, get ready," Stabler announced, but he could not make out the object in the man's grip. He instantly adjusted the scope's focus, but the man had bent down and was finished by the time Stabler zoomed in. Suddenly, a separate car that was parked on the opposite side of the road acknowledged the truck and they drove off in different directions. They vanished to the sound of Stabler swearing out loud.

He was furious. Everything had happened so quickly that he could not tell if the man had softened the ground or planted something. If he'd planted an IED, Stabler was responsible for any damage, or so he thought. Either way, it was over and he could only wait to see the outcome.

After minutes of cursing, Johnny relieved Stabler to get his mind off the incident, but Johnny was also thinking. As small as it was, their building was not an ideal defensive position. If his team made contact, they could easily be overwhelmed with well-placed RPGs and machine guns. Fallujah in 2004 had taught Johnny the effects of a well-trained enemy. Stabler and Johnny formed a plan.

"If we make contact, this is what happens," explained Johnny. "After I shoot, you all have to push out of this building.

I'll maintain contact until you all take cover, and then you guys open up. I'll take Governal and we'll sweep through the objective," he said.

They all agreed, and it was not long before the scene was thrown into action. Minutes later Johnny watched an eighteen-wheel semitruck stop where the other truck had been. Two men fell from the cab. One walked to the rear of the trailer while the other stopped directly center. Johnny had a waist-high view only.

"Hey, boys, we got a truck," said Johnny to his teammates.

The man near the center of the trailer caught Johnny's attention. He was squatting and reaching for something in the undercarriage.

"I got a guy. He's squatting," Johnny relayed to his team.

The man washed his hands from a spigot. When he was finished, he dug at the ground exactly where Stabler had described. Johnny watched closely but could not tell the size of the hole in the ground. He anxiously waited to see if the man was going to plant something. Suddenly the man pulled a blue jug from the undercarriage and placed it on the ground.

Earlier that year, insurgents had made IEDs with chlorine and nitroglycerine packed in jugs. Other reports from areas such as Adwaniyah confirmed that insurgents were using anti-freeze jugs for IEDs. Johnny considered the man's actions, combined with the actions of the men earlier, and the jug. The entire situation was similar to the procedures used by IED cells. Johnny could not let the man get away, and instantly chose his fate.

"He's got a blue jug," said Johnny, "I've got positive identification."

Stabler waited for Johnny's next move.

"So . . . ?" he questioned, wondering why Johnny was taking so long to shoot.

He did not know that Johnny was taking a breath to get his natural point of aim.

The suppressor fixed to the MK-11 helped mask Johnny's shot. The man did not know what hit him. Johnny's crosshairs lay on his chest, and the man twitched after the first shot, but Johnny did not let up. He fired two more times before the man stood up and stumbled away.

The team instantly executed the plan. Johnny moved the rifle to the side and grabbed his shotgun.

"Let's go!" he said.

Alex dropped the radio and picked up his rifle. Stabler followed Johnny through the front door while the SAW gunner and the rest of the team followed Alex, except one marine who stayed on the sniper rifle for over-watch.

Outside, Stabler led the team twenty meters to the front and took cover behind a few trees near a small dirt mound. Johnny and Governal moved up while the team laid covering fire.

The truck was still running, and by now others were getting out. Two of the men ran to the back of the truck, leading the marines to believe that they were going for weapons. Alex unloaded on a man looking toward him from behind a tire. Stabler lobbed a grenade from his M203, placing it under the truck and blowing out the back tires.

"Dump it!" shouted Stabler to the SAW gunner, to give Johnny and Governal suppressive fire. Stabler lobbed another grenade and landed it perfectly on the opposite side of the truck just as Johnny and Governal reached the road.

Johnny rounded the front of the truck with his shotgun resting firmly in his shoulder. He searched for any threats and noticed four wounded men on the ground. His eyes were drawn to one of the men reaching for a cell phone. He remembered that cell phones were used to trigger IEDs and the jug was only a few feet away.

"Stop! *Giff! Giff!*" he shouted in Arabic, but the man was unfazed.

Johnny did not hesitate, knowing that he and his team were dead if the man set off an IED. He leveled the shotgun and blasted the man in the face, immediately incapacitating him. Governal covered Johnny's back and warned a few cars to stay clear. He had turned in time to see the man crawling toward the phone. If Johnny had not shot him, he would have.

The remaining men raised their hands in surrender. One man leaned against the tires and looked toward Johnny.

"Insha'Allah," said the man with a smirk while he pointed to the sky. He meant that whatever happens, it is God's will.

"Who in their right mind would taunt marines at this point?" thought Johnny. Only insurgents with a death wish.

Vehicle traffic stopped in both directions as the marines tried reporting the incident, but the radio still did not work. A few hundred meters away, a four-vehicle convoy moved toward the snipers' position, having heard the shootings. Alex met them and tried to use a Humvee's radio to make contact with the COC, but the Humvee radio was not working either. He eventually turned to the satellite phone again.

Johnny hastily searched the truck. The men had no weapons, but then again IED cells typically did not carry small arms. With a glance in the cab, he moved to the back and blew the locks off the trailer. A few microchip boards and rags were strewn about, but not much else. He did not have time to search deeper into the trailer, nor did he want to, because the longer they stayed on the road, the more likely it was that they would be counterattacked.

All of the men on the ground had Syrian identification cards. There were also two Syrian-made cell phones and numerous religious tapes. Everyone suspected they were terrorists, and it was no secret that the prime route for foreign fighters infiltrating Iraq was from Syria.

The snipers loaded the QRF vehicles for extract while other marines cordoned off the area. Everyone was happy that they had stopped the IED cell, and though they found no weapons, the snipers expected the other marines to find IEDs or

IED-making materials when the truck was fully searched. The problem was that no one managed to completely inspect the truck.

When the snipers left the scene, the marines only opened the door to the trailer. Directly in front of the door were a few crates of soda, but no one bothered to crawl inside the fifty-three-foot trailer. Nor did anyone bother to inspect the undercarriage or the blue jug. Coincidently, a tow-truck driver, with no recommendation from the marines, arrived to drag the truck off. The marines did not treat the tow driver as suspicious at the time. Only later did they realize that he seemed out of place. They allowed the truck to try towing the eighteen-wheeler, but with all but one tire blown, it was impossible to move it. When the truck failed, it left, as did the marines.

When the snipers arrived at base, they were greeted with praise. A few recon marines along with other snipers heard about their success and congratulated them. Even the battalion sergeant major gave them a pat on the back. The mood quickly changed, however, when, after they cleaned up, they debriefed at the COC.

"Why couldn't you get comms with us?" yelled the intelligence officer after they walked in.

They did not have an answer. Only later, with suggestion from Alex, did they realize the problem. All Humvees in the battalion were equipped with IED jamming devices which sent an electronic signal a few hundred yards in each direction, essentially detonating any electronically controlled IEDs before the vehicles reached the bomb. They were great for their purpose, but the Humvees were parked so close to the COC that the jamming devices obstructed its radio frequency. It was a problem that went unsolved for some time.

The snipers were taken aback by the officer's temper. Johnny explained the incident, but the officer was furious. He was concerned because there were no weapons found. Stabler reminded him that the other marines were supposed to have inspected the scene, but they were already at base, also. It was too late in the day to inspect the truck now, and the command planned a thorough search the next morning. By then the truck was gone.

The snipers were further scolded.

"I don't know what happened, and I don't know what you guys saw, but I can't help you," finished the officer. He left a bad feeling with the snipers.

They knew that the officer had been investigated for the Haditha incident during their last deployment, and that he wanted nothing to do with this new incident, even if it meant throwing the snipers under the bus.

The circumstances of the shootings sparked controversy among the command. It did not matter that the Syrian men acted just as IED cells had, or that the truck had not been searched by the marines, the fault was put squarely on Johnny, and he was to be fried.

An investigation began on Johnny that evening. The

snipers were in lockdown and were not allowed privileges, but Johnny received the worst. The next day he was informed that his inability to make correct choices endangered the team. The result was immediate relief from his team leader position and expulsion from the sniper platoon. This was a sniper who had received meritorious promotions and awards, but all that was insignificant now. Everything that he had worked so hard for was taken in an instant, and it hurt.

For the rest of the deployment, Johnny was put on guard. Ironically, he was supposed to guard the people who wanted him prosecuted. Other marines steered clear of him, and the seriousness of it all hit Johnny after overhearing a gunnery sergeant tell his marines to keep away from the murderer, referring to him. To cap it all off, when the marines convoyed back to the main base to fly from Iraq to the USS *Bonhomme Richard*, one officer notified Johnny that he should not fire his weapon under any circumstance, even if they came under attack. When he heard that, Johnny knew right away that he needed a good lawyer.

A year later, the investigation came to a head. Johnny was taken to trial under an Article 32 hearing, similar to a civilian preliminary hearing, to determine if Johnny would face the most serious of court-martials, the general court-martial. The stress was enough to break anyone, but Johnny felt relieved to end a year-long wait and start the process.

The official trial lasted two days, in July 2008. The first day of trial was surreal. Johnny arrived at the small Camp Pendleton, California, courtroom in his service Charlie uniform, but the atmosphere was not as dramatic as he had anticipated. Everyone was cordial, and the prosecutor was nice enough to greet him with a handshake. It seemed that everyone genuinely wanted him to be found innocent, but the charges still proceeded.

He faced two counts of manslaughter and two counts of assault with a deadly weapon, with a minimum sentence of forty years in prison. A mix of emotions ran through him. He felt betrayed by the service he loved and angry that the officers responsible for the process were more concerned about their careers than about the men who fought for them. Deep down, though, he truly believed that he had done the right thing and that any marine, in his shoes, would have done the same.

In the courtroom, Johnny was not alone. His family sat behind him, with fellow marines and friends behind them. Local and national media representatives arrived to cover the trial. During the testimonies, the intelligence officer made him out to be a sloppy marine who had made a crucial error in failing to fully interpret the ROE. Johnny wondered what that paper pusher would have done had he been there, probably nothing but be paralyzed with fear. He did not hate the officer, but he felt bad that the man was willing to say anything to cover his own skin. It was obvious to everyone who really knew Johnny that he was anything but a sloppy Marine.

Machine Gun Sniper

Johnny was a Marine's marine. He had joined the Corps in 2003 after reading *Guns Up*, a book about a Marine machine gunner during Vietnam. After that, using a machine gun was all he wanted to do, and while at the School of Infantry, he became a 0332, Marine Corps machine gunner. When he graduated, he was sent to Kilo Company, Third Battalion, First Marines, who were in Kuwait waiting for the invasion of Iraq.

As big as he was, Johnny became the ammo bearer in his machine gun team during his first deployment. On the verge of combat, the marines in his weapons platoon welcomed him when he arrived, and though he was a boot, war was at hand, and he bypassed most of the games that new marines are run through.

When the war kicked off, Johnny was stuffed in the hull of an amtrack. He had just enough room to breathe next to the other marines, the weapons, and the ammo. His unit did not see much action during the first few days, but within a week they were fighting their way through the city of Nasiriyah.

The clash began when a U.S. Army supply truck took a wrong turn into the city. After realizing they were lost, the commanding officer retraced their route, only to be met by fighters waiting in ambush. Eleven soldiers were killed and six captured in the fight. Soon, tanks rescued the unit, followed by U.S. marines with an attack of their own.

Johnny joined the fight when his battalion was told to push into the city and hold security for the rest of the unit. Later that evening, Johnny's amtrack trailed a group of tanks across the Euphrates River Bridge and into Nasiriyah. They fought down a two-mile stretch of road appropriately named "Ambush Alley," which at the moment was controlled by remnants of Saddam Hussein's loyal militants, the Fedayeen. Johnny was wearing his NVGs while scanning the buildings when his amtrack was hit with small arms fire. The enemy looked like ninjas dressed in all black, running, firing, and moving from building to building. It was obvious that they had stashed weapons in different locations.

That night, Johnny helped fight off the attackers. It was his first firefight and very draining, but he was impressed with the massive amount of firepower on his side. M1 Abram battle tanks destroyed buildings, while .50-cal machine guns and MK-19 40mm grenade launchers mounted on the amtracks kept the enemy fighters off the streets. The next morning very little enemy resistance was left, but those who survived used inhuman tactics. Johnny was appalled when he watched some of the men fire in their direction while using women and children as shields.

The next day Johnny's unit was in full control. The fight in Ambush Alley lasted eight hours. Johnny was exhausted, but satisfied that he'd handled his first fight well. He had made contact with the enemy and lived.

Soon, his unit raced toward Baghdad. They cleared and bypassed nameless towns along the way. When they reached Baghdad, the marines received a surprising reception from civilians. Many Iraqis spoke adequate English and seemed to take interest in the marines' lives. Occasionally Johnny spoke with civilians about politics and other issues, and most of them were thankful for the U.S. presence.

A few weeks later, Johnny's time in Iraq was running out. One night he found himself next to a campfire listening to a few marines telling stories. At first he thought they were from his company, but soon he realized that they were snipers. From their stories, he understood that they had experienced an entirely different view of the war. Theirs was one from behind scopes, with vivid images of war and of kills they discussed knowingly. Right then Johnny wanted to become a sniper, but he would have to wait almost two years.

Trials

As the first day of trial came to an end, Johnny was in good spirits. The testimonies of his fellow snipers pointed toward the truth: his innocence. The courtroom examined the Rules of Engagement given to the snipers, and Johnny's platoon commander explained that senior commanders were not clear on when the snipers could engage and what exactly "positive identification" meant to the snipers The reality of it was that senior commanders had not entirely thought out a comprehensive explanation of ROE that was fitting for snipers. The dilemma was that snipers could prevent and disrupt enemy activities with positive identification of actions that appeared to have harmful intent. In other words, they could react based on observed actions that were similar to the enemy's MO, and Johnny had done just that.

That evening, when the proceeding had closed for the day, Johnny reflected on his time in the Marines. He still loved the Corps and wanted to serve, and the Marines needed warriors like Johnny, but sadly he'd experienced the dark side of the Corps, the political side. He stayed optimistic, and though this turn of events was unfortunate, it only added to his already long list of experiences. Of them all, none was more dangerous than his second deployment, where he had seen action in Fallujah in 2004.

On his second deployment, Johnny first settled in Al-Karma, Iraq. Tension between residents of this small city and the marines were felt immediately. The citizens held allegiance to the insurgency in Fallujah, Al-Karma's neighboring city to the south.

For the first part of his deployment Johnny was a regular rifleman patrolling streets and roads with the infantry. He arrived disappointed, wanting to be where all the action was, with marines from 1/5 and 2/1 in Fallujah. The incredible amount of fighting there is what drew Johnny. He wanted to help destroy the insurgency, but he would have to wait.

In one year, many things had changed in Iraq. Johnny's unit was on the defense when they arrived. It was totally different from their experience during the invasion, where they consistently set about offensive operations. Other things had changed, too. Now civilians despised U.S. troops and their presence, whereas during his first deployment, Johnny was welcomed in most places. Also, the fighting was more intense now, especially in the predominantly Sunni areas where Johnny happened to be. Above all, the major difference, though, was the insurgents' primary weapon, the IED, and it was not long until Johnny was affected by one.

One day his platoon set out to meet with a local leader. Security and stabilization operations pushed for marines and soldiers to uphold relations with locals. That day, Johnny's commander was to ensure that the leader was receiving U.S. supplies for his community. The meeting went as planned, and while Johnny waited outside for security, a group of kids threw rocks at his Humvee, a normal reception in that town. Johnny was glad to finish up, and the ride back to base was quiet. Almost no one drove on the roads, which should have been a sign. Johnny was standing in the turret on security when he glanced at the vehicle behind his. There he saw a gigantic plume of smoke forming into a mushroom cloud.

He immediately yelled for his driver to stop, while the

vehicle behind him was blasted with an IED. It flashed in his mind that he was supposed to be in that vehicle but had been moved at the last moment. Metal and dirt rained down around Johnny's gun truck before he and his interpreter ran for the injured marines behind them.

"Security!" yelled Johnny when he reached the battered vehicle. One side of the Humvee was completely bashed in, and the marines inside were severely injured. The first person he came across was Johnny's good friend. His cammies were shredded from the blast and his glasses lay shattered nearby. Blood oozed from his leg and he was passed out. Johnny cringed when he also saw that a chunk of the man's head was missing. Seconds later, the marine opened his eyes, but shock set in and he just stared into space. When he regained consciousness, he cried out for his brother, who was also a marine.

The platoon sergeant had also been hit but was fully conscious. He was missing both forearms, but he still directed the marines. The driver escaped with only minor injuries. For Johnny, the destructive power of the IEDs left a lasting impression.

Though this was his first brush with an IED, it was not his last. Weeks later, Johnny was manning an entry control point in a marketplace in town. Interaction with civilians was always exciting, but also dangerous, especially while remaining in the same place. That was the best way to get ambushed or mortared. That day a man sparked up a conversation with Johnny.

The Arab spoke broken English and talked of his home country, Jordan, and was very sociable, asking Johnny about politics and economic issues. The conversation was short and more
enjoyable than being watched suspiciously by the locals.

After the man left, Johnny noticed something on the ground
behind him. It looked suspicious, and he casually examined the
item, hoping to avoid attention. When he stepped closer, he
realized that it was a mortar shaped into an IED. He tried not
to panic while walking to his squad leader nearby and telling
him the news. His squad leader was skeptical at first, but soon
he confirmed that it was an IED. They cleared the area without
the device exploding. That area was notorious for IEDs, which
had previously injured three other marines there. The battalion
wanted no more of that and allowed the squad to try stopping
the men placing bombs there.

Johnny was furious. The man had set him up. He did not
dwell on it too long, because that night his squad set up an
ambush. At 0100, they patrolled through the streets and alleys
near the previous spot, and surprisingly the entire squad made
it into position without being compromised. They hid in a
palm grove and spread out on watch.

With his Aimpoint, Johnny scanned the roadway. The
bipods under his M16 steadied the rifle while he placed the red
dot reticle on target areas. Hours later a man appeared 150
meters away. Johnny focused his night vision and saw that the
man was digging on the road. No one in their right mind

would be digging on the roads in the middle of the night, especially because the ROE stated that anyone doing so would be shot. That was all that Johnny needed. He had positive identification. The target fit the ROE, and he took aim.

When the man stopped digging, Johnny placed his reticle on his back and fired two shots. The man fell right there. Seconds later a car appeared from the shadows, loaded the injured man, and drove off. Johnny convinced the squad leader to search the area and found blood next to a small hole where the man had been digging. Consequently, after the incident, there were not many IEDs emplaced on that stretch of road.

The next few months were uneventful for Johnny, except for the occasional house-to-house search. All the while, the situation in Fallujah was getting more intense. The marines from 3/1 heard that they might be going over to help with fighting the insurgency, but they had to wait until after summer, when the temperature dropped.

By October 2004, Johnny and his unit had gotten the call to be part of Operation Phantom Fury. The offensive mission called for Marine and certain Army units to clear the Sunni stronghold. The men in Johnny's unit were sick of the patrolling they had been doing for months and that they were frequently targeted with IEDs. To add to their frustration, the insurgents rarely dared face Johnny's unit head-on, but this was marines' time to get some payback.

Every marine in country wanted to be in the fight, and

Johnny was no exception. Fallujah was said to be in complete control by insurgents, who declared that nobody would take it from them. By November that declaration was going to be put to the test.

Days before the operation, Johnny was retasked. His machine gun team was given a Humvee. Their new gun truck had two M240G light machine guns, and Johnny decided to use one for reserve. His truck was to operate in conjunction with another Humvee mounted with an MK-19 fully automatic grenade launcher. They would support each other while providing fire for the advancing infantry during the fight. Together, the supporting vehicles and the infantrymen prepared for the fight and rehearsed the action to come.

Inside the city, insurgents had plenty of time to prepare, as well. They had held the city for more than six months, with no U.S. patrols entering in. A few thousand enemy fighters organized ambush sites and prepared weapons and ammo caches, as well as booby traps. They promised to counterattack once the marines set foot in the city. They received warning from the marines, and by the time the attack began, the insurgents, not willing to fight to the death, were gone.

On November 8, 2004, U.S. forces began their assault. Johnny realized he was bound for heavy fighting when an Army Special Forces ODA team and a Navy SEAL team were attached to his company. That first morning, his battalion cleared the train station north of the city and waited for a breach on the railroad tracks by the Army's Second Battalion, Seventh Cavalry. When the hole was made, marines advanced behind the armored cavalry.

The northern edge of the city had a cluster of buildings, roads, and small alleyways. Johnny was standing next to a friend in the turret when they moved in. They would be side by side through it all, Johnny with his M16, while the other marine gripped the machine gun. Insurgents weaved through buildings and alleys, waiting to attack. By the time Johnny entered the city, the fighting was intense.

His first contact came while moving into a narrow alleyway. Farther down, insurgents appeared and launched RPGs at the marines, who in return pressed toward them, weapons blazing. Johnny emptied a magazine, but with so many marines shooting, he could not tell if he had hit anyone. Surprisingly he had not been hit. Bullets sent sparks next to him.

In no time, the marines discovered the insurgents' tactics. Their goal was to draw the marines' fire in one direction using AKs and machine guns. It was a diversion for RPG gunners to fire their rockets from other directions. The marines countered using their vehicles to bound ahead of the infantry and hold security while the dismounts moved. Johnny's truck and the others pushed forward and held in position, allowing the marines on the ground to catch up.

The farther the marines pushed, the more fierce the fighting felt. Once, Johnny engaged insurgents as close as fifty yards away. His M16 was more maneuverable than the M240G; he wounded the enemy and his gunner finished him off. Johnny could see and smell smoke, and gunpowder stained the air. In all the shooting, he could feel the heat from the barrel of his rifle. Once, he noticed a SAW gunner next to his truck dashing to a corner, but just as he was about to bypass the alley, he was fatally wounded. The marine was Johnny's friend he had known since his time at the school of infantry. With death so close, Johnny used it as motivation to fight harder.

Hours into the fight, Johnny learned that most of the snipers in his battalion had been injured. The SEAL and Army ODA snipers were brought up in their place, and they began making quick work of insurgents in the open. From elevated positions, they were able to finish off wounded insurgents and eliminate others in the distance. In the heat of the battle, Johnny noticed an insurgent with a red head wrap run into the street.

"Is this guy serious?" thought Johnny, as the man fired in the open with a machine gun on his hip. His aim was horrible but his intent was obvious, and Johnny shot back, killing him. All around them, the marines were astounded by the fighting of insurgents. Reports of enemy fighters being drugged were true, and often, though severely wounded, the insurgents were able to drag themselves away before dying.

That day, death was closer to Johnny than he knew. While he was in the turret, RPGs narrowly missed him and the machine gunner. Once, Johnny was shooting, when the machine gunner next to him suddenly ducked into the Humvee.

"Get back on the gun!" Johnny yelled, noticing a plume of smoke close by.

The wide-eyed gunner stood up behind the machine gun and resumed fighting. Only later did he explain to Johnny that he had ducked because an RPG skipped off the turret's deflection shield. Had it exploded, they would have both been dead.

By the end of that day, the marines had pushed up deep into Fallujah. Air support kept insurgents at bay, while U.S. snipers picked off others. Johnny's company set up in defensive positions and found writings on walls.

"Your armored vehicles are just target practice for us," the writings warned, implying that insurgents took pride in disabling and killing those inside vehicles. Unfortunately, that day the insurgents were able to destroy two M1 Abram tanks with RPGs.

Though the first few days were hairy, Johnny's most memorable experience in Fallujah happened later. Near the end of the assault, as the marines cleared up the last pockets of resistance, two platoons became bogged down in a heavy fight near Johnny's squad. His trucks were called for support and they raced to the scene.

When they arrived, they found one alley completely controlled by insurgents with machine guns and RPGs. No vehicles dared move down it.

Johnny's squad dismounted to help suppress the alley. Usually AT-4s were reserved as anti-armor weapons, but in Fallujah they were great for breaching walls or hitting houses occupied by insurgents. The marines were out of rockets, but Johnny overheard someone asking if anyone knew how to use the RPGs found on the streets.

"I know how," lied Johnny. He'd never fired the weapon before, but he had researched it online. His platoon commander gave him the thumbs-up, and he took a bundle of rockets into the alleyway. One marine opened with suppression fire and Johnny turned the corner to unload. He fired a few rockets, but was not able to see the impacts. The marine next to him said that he sent one directly into a window where insurgents had been spotted. Johnny's rockets had stopped the enemy fire.

The platoon commander was impressed. He directed Johnny to fire on another nearby house filled with insurgents. Johnny and others charged the building and ran through its courtyard. When they entered the courtyard, a dead insurgent lay at their feet. He gripped a hand grenade and wore an American flak jacket with a chest harness full of AK magazines. It struck Johnny just how fanatical the man was about killing the marines. He fought to his dying breath, while still clutching a hand grenade that had no pin in it.

Johnny stepped over the man and moved to within feet of the front door. If anyone was behind the door, they were killed instantly when Johnny sent an RPG through it. The rocket also hit a fuel drum inside, igniting the entire building. Slowly fuel began to leak under the doorway, sparking a fire in the courtyard. Johnny and the others ran back to the squad, exhausted from the fighting.

The next day, the marines were near their final objective in the city. They had pushed all the way through and killed or captured most of the fighters, but they needed to clear the last of a few buildings near their firm base. Johnny's vehicle held security for infantrymen clearing houses. He trailed a squad of marines to a home outside the base. As soon as the marines entered, gunshots rang out. A SAW went off, answered by AK fire, followed by M16s. Johnny knew right away that the house was packed with insurgents and that the marines were in for a fight there.

Inside, the marines moved to clear the bottom floor. The point man spotted an insurgent squatting and shot him dead. Next, the point man entered another room and was met by another insurgent. The two exchanged fire from a close distance, and the marine was able to unload a long burst into the insurgent's chest, causing his clothing to catch fire. Surprisingly, the insurgent was still able to crawl toward the marine before dying. When the lower floor had been cleared, the marines pressed upstairs. Unbeknownst to them, a group of insurgents had been alerted by the shooting and prepared a counterattack. When the marines made it onto the stairway, the insurgents opened fire with AKs, machine guns, and hand grenades.

A few marines were instantly injured. The insurgents had the advantage, pinning the marines inside and preventing anyone from entering through the doorways. Though injured, the marines crawled into rooms on the bottom level. First Sergeant Brad Kasal was one of the first marines to rush in and help, but he was instantly cut down. He pulled another marine into a room and held off the attackers while shielding his comrade from grenades with his own body.

Outside, insurgents dropped grenades from the second level, isolating the entrances. Johnny desperately wanted to help after learning about the casualties, but he could not attack the building and risk hitting his fellow marines inside. At first break, Johnny dashed inside and met up with other marines being held at bay next to a doorway. A dead insurgent lay feet way, his hands still gripping his AK. Johnny moved to the doorway and poked his head around the corner to glance into the next room.

A smoky haze smoldered in the room. Johnny saw injured marines on the ground and heard their faint calls for help. Suddenly, a burst of enemy fire blew toward him and bullets cracked inches from his face and hit the wall beside him. He would not do that again.

For the next half hour Johnny and the other marines

exchanged fire with the insurgents from the doorway. They desperately wanted to evacuate the wounded, but there was no way to get into the room without being shot. Everyone feared that the marines would bleed to death, and the marine next to Johnny began to get desperate. He began to weave in and out of the doorway to fire on the insurgents, but he was playing with his life. Finally, the marine exposed himself for too long and was shot three times in the head. He died right there next to Johnny.

Finally another platoon arrived for backup. Johnny went outside to talk with the lieutenant. The lieutenant wanted to pressure the insurgents. He ran across an alley to get a better perspective on the second level and then began throwing grenades into the windows. Johnny went back inside and was putting fire through the doorway when a bullet grazed his knuckles.

The insurgents began to run low on ammo. Minutes later, firing from the insurgents' position let up, allowing Johnny and another marine to run through the doorway into the other room. Inside, the other marine helped the wounded while Johnny held security. They did not want to risk moving the wounded through the doorways and yelled for a Humvee to move close. They attached a pulley to the security bars on the windows and the vehicle ripped them out. The opening allowed them to lift the wounded out of the house.

Meanwhile, other marines flooded the house as well. They were able to evacuate everyone from the bottom floor. Once outside, the marines threw a satchel charge into the house and destroyed the entire building. It caved in on itself, and Johnny saw blood spray from the explosion. When everything settled, Johnny persuaded the platoon commander to search the rubble. They scoured the dirt and concrete and came across a heavyset man, half-buried in rubble. He looked dead at first, but he started to move when the marines closed in. With one arm, he retrieved a grenade from his chest harness, pulled the pin, and rolled it in the direction of the marines.

Everyone scattered, but the grenade exploded next to the insurgent. Unbelievably, the man moved again, reaching for his chest harness. The entire platoon fired at him but did not stop him. The insurgent was definitely drugged. The platoon commander walked up to him and dumped fifteen rounds into his chest. Incredibly, he did not stop reaching for his chest harness. Johnny had had enough and finished him off with three thrusts from his bayonet.

When the battle was over, Johnny was thankful to be alive. Though they had won, it came at the price of the health and lives of many of his friends. He gained a new respect for the enemy, and it changed his outlook on their capabilities. He would never again underestimate their ability to adapt to U.S. tactics or their tenacity to fight to the death.

When he returned home from Fallujah, Johnny tried out

for his battalion's sniper platoon. The conditions were miserable, but he stuck it out and was selected along with seven others from sixty marines. He went on to do two deployments in the sniper platoon, the second being the deployment to the Lake Tharthar region.

After the second day of his trial, Johnny felt there was a good chance of the charges being dropped. Despite the media's opinion, the testimonies of his teammates showed his true intentions. They all trusted and believed that he had done the right thing. The convening authority, a judge advocate general, considered that Johnny had been through the fiercest of fighting many times over and never lost his bearing, nor had he done anything questionable before. Johnny's attorney knew that there was not enough conclusive evidence to push the case for a further trial.

The next day, Johnny's dad called him and told him that the charges had been dropped. The judge concluded that the trial should have been handled at the battalion level, and after review by the First Marine Expeditionary Force general, he concluded that Johnny had acted with reasonable force. He recommended lesser charges, but did not dismiss the case entirely. It was dropped without prejudice, meaning that if further evidence was presented, the case could proceed.

Relieved, Johnny felt the ending was bittersweet. The least the command could have done was to call him and tell him the situation. It was somewhat disrespectful that he had to hear the results of the trial from his dad, who had heard the news over the radio. Regardless, the nightmare was finally over, and Johnny could do what he'd told the judge at the end of the trial:

"I'm eager to actually get back in the fight and serve my country."

ELEVEN

COST OF WAR

WAR changes lives. Fighting, killing, and dying are experiences that those who live through them will never forget. For snipers, the battlefield and all it encompasses are witnessed at a magnified level. Every engagement, every kill and attack, is seen up close and personal, and the psychological trauma associated with this affects everyone in different ways. These days some snipers return from war only to fight a new battle at home in the form of an anxiety disorder known as Post-Traumatic Stress Disorder (PTSD). As Marine Sniper Sergeant Byron Hancock knows, this disorder can be crippling, but it isn't unbeatable.

The War

Byron was well prepared for combat. He joined the Marines in 1988 as an infantryman in the reserves. His uncle and grandfather were both marines, and being a proud Texan, Byron wanted to serve as one, also. When he began the school of infantry, one of his instructors had been a sniper, and when Byron showed interest, the instructor told him everything he knew about sniping. That set Bryon on a path to becoming a sniper himself.

As a weekend warrior, he checked into his unit, the First Battalion, Twenty-third Marines, a reserve battalion in Texas. Soon he was part of the sniper platoon, and when he had three months of free time, he was sent to sniper school in Quantico, Virginia. It was a tough class, but Byron graduated and became a hunter of gunmen in 1991. In the next ten years he stayed with the battalion, but he was on the verge of discharging when the events of 9/11 made him reenlist. Three years later he was shipped to Iraq.

He arrived to the Al Anbar Province in 2004, well aware that as a sniper, he was going to face the insurgency. By then, his fourteen years of service made him the sniper platoon's chief scout/sniper. Weeks after arriving, Byron began sniper operations, and a few months later he and his partner took part in Operation Phantom Fury in Fallujah.

In his civilian life, Byron was a cop. A peace officer, to be exact, and he patrolled the town of Bryant, Texas. His wife back home cared for their four kids; she was his high school sweetheart and he loved her more than anything. She was his reason to live through Iraq.

At thirty-four years old, Byron was significantly older than his partner, Flowers, who was in his early twenties. When they learned about Fallujah, they were excited at the possibility of action. The two of them joined Bravo Company and fell under Regimental Combat Team Seven. They were to hold the peninsula on the west side of the Euphrates River bordering Fallujah. The main force would clear the city while Byron's element cleared the small villages on their side of the river to keep insurgents from escaping that way.

Once they reached Fallujah, Byron felt the intensity right away. For weapons, he had an M40A3 and a pistol. Some snipers might have wanted an M16, but Byron was comfortable with a pistol. As a nine-year veteran of the police force, he was an expert with it. His partner, Flowers, had an M16 with an M203 grenade launcher. The two of them were in direct support of Bravo Company. Their time there clung with Byron even long after leaving the country.

Gunfights broke out from the very beginning. IEDs hit patrols and mortars rained down on the marines for the first few days. It kept Byron on edge the entire time. Soon his unit began to take casualties, and it made Byron even more alert. He knew that if he let his guard down, he could die at any moment.

After days of patrols and over-watching, Byron and Flowers rested at their unit's command operations center. They were exhausted from constant operations. On their first day back to rest, their temporary base took mortars. The explosions injured a few marines, and the snipers were quickly directed to find the insurgents responsible.

Early the next morning, Byron and Flowers patrolled from their base. The two of them skirted tree lines and fields while making their way down the peninsula. As they passed through buildings, Byron pulled his pistol and cleared rooms with it, and soon they were on a rooftop. Smoke billowed from across the river, as other units battled insurgents. Byron and Flowers had their own fight, though. They needed to find the mortar men.

Reports indicated that the snipers were in the area where the mortars originated. By morning Byron scoured one sector while Flowers searched another with his spotting scope. Their rooftop allowed them to see great distances in every direction. Byron kept his sights on a tree line while Flowers focused on a certain group of buildings. Though it was tedious work, their patience paid off.

"Hey, I've got five guys over here," Flowers said to Byron.

Byron turned his attention to the men. His Unertal 10-power scope revealed them standing next to a building. A few men pulled a mortar tube from nearby and began setting it up. Quickly Byron told Flowers to start a fire mission with 60mm mortars. When Flowers got on the radio, Byron found the distance to the men.

The laser range finder put the men at 1,050 meters (3,500 feet). Byron fixed the elevation on his scope and watched the area to make a wind adjustment. There was none, and he took aim. With the men in sight, he stayed patient waiting for the first round of the mortars to drop. It would keep their position concealed.

As he waited, Byron concentrated on his breathing and the crosshairs. At that distance, if he was off even just a slight bit, his bullet would not impact the target. Within minutes, the mortars were on their way, and Byron was ready to shoot. Flowers sat behind the spotting scope ready to inform Byron of his adjustments should he miss.

The group of men huddled next to a building surrounded by knee-high grass. With his crosshairs, Byron aimed at a man standing. He was clearly the best target, giving Byron the entire view of his back. Moments later, the first mortar round splashed, but it was off target. Byron, however, was not. He squeezed the trigger and Flowers watched the man fall.

"Hit," exclaimed Flowers.

When his rifle settled, Byron didn't have to move his crosshairs. Another man who had been squatting stood up right where the other man had been. Byron could not have been luckier; he quickly settled his sights and fired once more.

"Hit," said Flowers again. At the same time Flowers radioed

the adjustments for the mortars, and the next round was dead-on. After Flowers called in fire for effects, the next few mortars impacted near the men, engulfing them in a cloud of dust and killing all of them. Byron didn't think much of the kills; they weren't his first, but at the time they constituted the longest shot recorded by a sniper in Iraq.

With the men killed, the command operations center was not mortared again. Byron and Flowers were heroes for a day before they were ordered out on more operations. They moved with Bravo Company to clear more villages. Meanwhile, across the river the fight was on. Marines and insurgents battled for days, all within sight of Byron. He watched tanks and air assets destroy everything. Bombings continued day and night, and it was an incredible sight to see. Once Byron and Flowers watched an AH-1H Super Cobra get hit in midair. The helicopter hovered 200 meters (650 feet) off the ground. Suddenly a rocket struck the side of it, sending it into a spin, but fortunately it landed safely.

When the fight in Fallujah ended, the marines from Regimental Combat Team Seven stuck around. Insurgents moved about on their side of the river, and for months Byron and Flowers helped with raids and patrols. Often, they set out on their own to observe villages for enemy activity and keep eyes on the roads to prevent IEDs.

One night, the two snipers were buried in a hide outside of a small village. Days earlier they had provided over-watch for marines conducting a raid there. Byron wanted to watch the area, knowing that further operations were planned there. The snipers focused on a single bridge, which crossed a canal. It was the only way into the nearby village, and if any IEDs were to be planted, Byron figured that along that stretch of road was the most likely region.

From their hide, the snipers spotted something. Byron noticed a car stopped on the bridge, but even with night vision it was too dark to tell what the occupants were doing. The ROEs wouldn't allow the snipers to kill without positive identification, and the car quickly disappeared. They reported the situation and waited to see if the men would come back, and the next morning, the car showed again.

Again, the vehicle stopped on the road. Byron and Flowers were completely hidden four hundred meters away. Byron watched intently through his scope waiting for the occupants' next move. Suddenly, a passenger jumped from the car and made his way to the side of the road. He wore a white dishdasha, known as a man-dress, and slippers and bent down. Under the watch of the snipers, the man grabbed wires leading from a bush. Wires were used for IEDs, and that was all the snipers needed to engage.

"That is the tallest Iraqi I've ever seen," thought Byron. He was well over six feet tall. Byron set his crosshairs on the man and fired a shot, hitting him in the chest. He tumbled backward and out of the snipers' vision. Next to Byron, Flowers opened up on the car and killed another passenger just before the car sped off. When the shooting stopped, both snipers were still full of adrenaline, but they knew that they needed to leave before more insurgents appeared, and they called for extract.

The next night, the team slipped back into the area. Under darkness they moved up to the crossing to see what the man had been reaching for in the bushes. They found the wires leading into the dirt. Byron and Flowers hoped to catch more insurgents checking on the wires, so they found another position and dug in. The next day a man on a motorcycle passed through the area slowly but did not stop. It was obvious that he had looked into the bushes where the wires were, but he'd had enough sense to keep moving.

Later, Flowers radioed the company. He explained what they had found, and soon an explosive ordnance disposal team arrived. The wires led to three 155mm artillery rounds buried in the bridge. The marines destroyed the explosives and moved from the area, taking the snipers with them.

Days later, Byron and Flowers split up. They were given new partners so that the company had two sniper teams available. Byron was dead tired from running missions, but he still carried on. Byron couldn't shake the thought of the man on the motorcycle and wondered what he had been doing. He convinced his commander to let him observe for five days the same area where they had found the IEDs. He suspected that the man on the motorcycle was up to something. The commander agreed, and Byron was sent out again.

He and his teammate spent an entire night patrolling and digging in. They were settled by dawn, and for five days they watched the man on the motorcycle drive to and from the bridge, but he never stopped. He only slowed down and looked at the bushes on the side of the road. That drove Byron crazy; he didn't have enough evidence to kill the man, but they knew he was part of an IED cell. Finally, though, on the fifth day, the man returned, but this time he felt comfortable and stopped on the side of the road.

From their hide, the two marines watched the man get off his motorcycle and walk to the nearby bushes. When he bent over, Byron saw him grab a set of wires. In a flash Byron moved his sights to the man's motorcycle and waited for him to get back on. It gave him enough time to settle his rifle. Moments later the man reached his bike but was met with a bullet from Byron. The bullet caused the man to stumble, but to the two marines' surprise, he hit the throttle and began to drive away. It was apparent he was hit in the midsection, as he covered it with one hand and leaned forward while driving. Byron aimed again, but the man moved too quickly. The two marines wondered how far he actually drove, because he was off balance and could barely handle the motorcycle.

When he was out of sight, Byron's partner radioed the

situation. Explosive ordnance technicians arrived and were guided onto the wires, which led to more IEDs three hundred yards from the overpass. Later it was discovered that the insurgents planned to blow the IEDs on the bridge as soon as a patrol passed over. They anticipated that the marines would retreat, and when they did, they were to be met with three more IEDs. Fortunately, Byron, Flowers, and the other marine was able to stop it from happening.

For the next few months, Byron ran missions near Fallujah. He learned that concealment was his only advantage. His two-man team dug into different dirt hides to blend in. They stayed off the beaten path, knowing that shepherds informed the mujahedeen about the Americans' whereabouts. By early 2005, Byron's stint in Fallujah was over, and his unit convoyed north to the city of Hit just in time for another operation.

Operation River Blitz was where Byron made his most memorable shot. The unit was tasked with clearing the city of Hit. The city was a hotbed for insurgents. It had not been patrolled, and by the time Operation River Blitz was under way, the marines expected a fight.

Days into the operation, Byron and his spotter lay on a rooftop observing south of their position. Soon single gunshots cracked nearby; someone was shooting at the marines on the ground. Earlier, one marine reported that he had been standing near a pole and felt the urge to step forward. Just after he did, a single shot hit the pole where his head would have been had he not moved. It definitely indicated that an enemy sniper was in the area.

When he heard the shots, Byron grabbed his rifle. He had been looking for the sniper for some time and kept his sights on the area where he would have hidden if he were the enemy. It was toward the town's ruins, a jumbled collection of fallen buildings and loose debris. A minute had passed when Byron had an idea. He told his spotter to move to the far side of the rooftop while he took to the other side. His spotter raised his helmet above their wall, and within a few minutes, somebody took a shot at it.

The thump of the enemy rifle drew Byron's attention. He caught sight of a man crawling from a hole within the ruins, but the man did not have a weapon. It was a common tactic by insurgents to shoot and leave their weapons behind, knowing that Marines could not kill unarmed individuals without probable cause. Byron kept his eyes on the area to see if the man would return. For the rest of the day and during the night he waited, scanning the area where the man had crawled from.

Before sunrise, Byron and his partner moved to an elevated position. They had ranged the area of the ruins, and he set his sights at 625 meters (a little more than 2,000 feet). The next morning, just after sunrise, Byron saw a slight reflection shining from under a pile of rubble. He dropped his binoculars and reached for his rifle. His scope revealed that the glint was the reflection of an SVD-Dragunov, and the man behind it sported a mustache.

With his target acquired, Byron told his spotter and they prepared for the shot. When he was ready, Byron fired once, and it was all that was needed. The man was hit in the head. Thirty minutes later six others arrived to the area where the man had been killed. Byron moved positions and could not see exactly what they were doing, but they were not armed. Other marines reported that they carried the dead man's body away and laid him next to the road.

In seven months, Byron had experienced the war in Iraq and all that it had to offer. Underneath, he knew that at any moment his life could have been taken. To survive it all, a certain amount of paranoia was needed along with the ability to kill. The weight of it all, everything that he had done and seen, was tucked away to be dealt with at another time. Iraq was no place to ponder his circumstances, especially with his life on the line. Little did he know, however, that it would all catch up with him soon enough.

At Home

By late March 2005, Byron was home in America. In a matter of days he went from a war zone to civilian life, and the adjustment was difficult. He returned to his wife and family in Texas, and all was well for the first month. As the days passed, however, slowly his wife noticed a change in her husband. Nightmares started in, and Byron often woke up believing that he was still in Iraq. Not only were his sleeping habits changing, his mind began changing as well.

He no longer enjoyed the things that he used to. Spending time with his family wasn't a priority, and he turned to drinking alcohol heavily. To make it worse, the alcohol triggered fits of anger and paranoia. Byron didn't know how to deal with the images in his head, but he didn't think that there was anything wrong and continued with his everyday life.

Soon things got worse. Simple tasks such as walking down the street and driving on roads were difficult for Byron. In Iraq these would have put snipers in a vulnerable position. Byron also couldn't stand not having his household secured, and he always locked the doors. When he returned to work as a police officer, he discovered that his behavior had changed and he reacted differently than he would have before Iraq. He did not take kindly to stupid questions, nor did he have the patience to deal with irritating civilians, and one scenario in particular made him realize that he had a problem.

It came when he responded to a call about a man barricaded in a house with a gun. When Byron pulled up outside, he immediately felt as though he were back in Iraq. The Rules of Engagement were different, but Byron still drew his pistol and waited for the man to show himself. In his mind, Byron knew that he might have to kill again, and in fact he was dead set on it.

The man inside unknowingly walked a fine line. He yelled threats at the officers and moved in and out of windows. Soon he showed himself. The man set foot out on his porch with the pistol in his hand, but he held it lowered by his waist. The job of the officers was to defuse the situation peacefully, but Byron took aim at the man's head. In Iraq, he would have killed the man by now putting an end to the situation with one bullet.

As he started to squeeze the trigger, however, something inside him told him to stop. This wasn't the war. Eventually the man surrendered, but the situation frightened Byron. To him, the value of human life did not carry as much weight as it had before. He'd been seconds away from killing, and he realized that he did not care if he had to or not.

At the same time, his home life was getting worse. Certain smells, sights, and actions triggered flashbacks of Iraq. He began to see his kills, and the scenes of war. He began not caring about anyone or anything, and the drinking got worse. His constant isolation was his way of not having to talk about the war. Soon, though, all he wanted to do was stay locked up in his house. His wife asked why he didn't want to leave. She wondered why he was so angry all the time, and why he didn't care about anything. Byron didn't know that he was suffering from post-traumatic stress disorder.

The anxiety disorder had hit him like a ton of bricks. The cause of his problems stemmed from the actions and events he'd seen in combat. The constant paranoia needed to survive in Iraq, along with being attacked and killing, had done things to Byron's mind. He wasn't even aware that he had the disorder. Fortunately for him, his wife recognized the symptoms and wanted him to deal with the problem, but it took some convincing. Byron reluctantly sought help with the local Veterans Administration office. He learned that he was definitely affected by the disorder, and that others suffered from the same problem. It was hard for Byron to admit that he had a problem. He did not want to be labeled, but soon he was admitted to a program.

When he returned to the police force, Byron was met with bad news. They had learned that he had PTSD and were concerned with him carrying a gun. They held a conference and talked with Byron about his position. He knew where they were going. Byron had been a cop for eleven years and received numerous awards, some even after returning from Iraq. In the end, however, he knew he would rather resign than be forced out, and that's just what he did.

Overall, Byron is learning to cope. Just like thousands of troops returning from Iraq and Afghanistan, he found that adjusting to PTSD takes hard work. He has learned to deal with his problems with the help of his family and the PTSD program offered by the Veterans Administration:

I've learned to let go and move on. I've learned to focus on what is important in my life and to enjoy doing that. Talking with other veterans and my family helps a lot. Finally, putting the war behind me and moving on is what has helped me the most. Figuring out how to do that is different for everyone, but that is the key.

TWELVE

BEYOND

BEYOND conventional war fighting, the need for snipers is expanding at every turn. Contingencies across the globe call for the skill provided by snipers. From civilian to military operations, there is no doubting the precision that trained snipers and their weapons bring to the fight. If the situation is critical, most often snipers will be called upon.

On April, 8, 2009, Somali pirates attempted to hijack the United States–based container ship, the *Maersk-Alabama*. The ship was filled with humanitarian supplies and traveled a few hundred miles off the coast of Somalia. Aboard, twenty crew members went about their daily activities well aware that pirates patrolled the area, but they did not expect to run into any. Soon, however, pirates appeared on the horizon and sped toward their ship.

Four days later, three Somali pirates drifted in the Indian Ocean with the captain of the *Maersk-Alabama* as their hostage. Their hijacking attempt had failed, but they were able to escape aboard a lifeboat from the ship, with the captain. A United States guided missile destroyer, the USS *Bainbridge* and another warship, the USS *Halyburton*, were dispatched to the scene. Later, one pirate boarded the *Bainbridge* after being convinced that he needed medical attention from a wound sustained in a struggle with the *Maersk-Alabama* crew. The three others, however, kept their weapons fixed on bound Captain Richard Phillips. Refusing to surrender, the pirates forced a standoff and brought about their own destruction when a U.S. Navy SEAL team was directed to the scene.

Under darkness, the Navy SEALs infiltrated the area by parachute. Along with their weapons and boats, they jumped from a cargo plane. Once in the ocean, the commandos made it to and boarded the *Bainbridge*. Among the team, three SEALs carried sniper rifles, presumably MK-13 bolt-action .300 Winchester Magnums. These weapons were preferred for their heavy barrels and reliable accuracy.

Aboard the *Bainbridge*, the SEALs took action. These men were most likely elite even among Navy SEALs and part of DEVGRU, or the United States Naval Special Warfare Development Group. SEAL Team Six, as they were known, were prepared for just such missions, and they needed no guidance or direction. Three of the men were snipers and occupied different positions on the ship's fantail. Elevation allowed them to look down on the lifeboat, which had been tied to the *Bainbridge*, allowing the pirates a smoother ride in the destroyer's wake. After sundown, the snipers attached PVS/22 universal night sights to their rifles while their spotters guided them onto their targets. The ships provided enough ambient light for the snipers to see their targets well.

During the night the lifeboat attached to the *Bainbridge* was quietly pulled to within thirty meters of the ship. The snipers had each designated certain targets for themselves, and kept their crosshairs steadily on them. Through their scopes, the pirates' heads would have fully encompassed the sights, allowing them to hit their targets anywhere in the head if need be. Their weapons, capable of holding three shots within a one-inch-by-one-inch target at one hundred yards, could easily score instant incapacitation kills, leaving no time for physical reactions from the pirates.

As directed by their commanders, the snipers were to strike when the captain was in imminent danger. Behind their rifles, the snipers had the confidence needed to make the shots. Their training had encompassed such scenarios, and they had a few things in their favor. The pirates were not aware that the snipers were tracking their every move. Despite the small boat moving in such seas, with their rifles and optics, the distance made the shots incredibly easy for the snipers. The only thing they waited for was for the targets to expose themselves enough that the snipers could fire simultaneously.

Suddenly, two pirates appeared from an opening in the life-boat. The third man held an AK-47 to Captain Phillips. It was time for the snipers to react. Each SEAL sighted in and held a target. A quick countdown allowed them to fire at the same moment. In a flash, three bullets ended the hostage situation.

The situation was just the type of scenario that calls for snipers. It's no surprise that trained professional snipers are now being heavily sought after in the military. Just as the military calls upon snipers, however, so also does the civilian sector.

Unfortunately 9/11 has had effects on security, both nationally and internationally. However, as a result, new opportunities are available for snipers who decide that the military is no longer an option. Now security contracting companies providing high-level protection see the need to incorporate snipers into their workforce. Companies such as Triple Canopy, Xe Services (formerly Blackwater), and DynCorp International are just a few where snipers can find employment.

As hazardous as it may seem to some, working for a security contracting company has its upsides. Phenomenal pay, freedom, and the ability to use one's skills are draws to these positions for snipers. But while providing some benefits, security contracting jobs *are* dangerous, whether you're a sniper or not.

AJ, a former marine, has served in Iraq as a military sniper and a contractor sniper. He explains the details of working for the Department of Defense as a security contractor:

I served in the Marines, and spent most of my time in a reconnaissance battalion. There, sniping is usually a secondary occupation until the skill is needed during war. As a result, I exited the military to be a government contractor, not specifically to be a sniper. I started off doing personal security details, and once the company I worked for found out that I had the specialty of a sniper, they wanted me to become certified as a State Department sniper. The extra certification pays more, so of course I jumped at the chance. In this industry, keeping a job is very competitive, and the more credentials you have, the better the chance that you will keep working.

When asked about the missions and the benefits of contractor sniping, AJ candidly replied:

The missions are very simple. Once the team receives their principle (the person that they're designated to protect), then mission planning starts. These principles are government officials and other VIPs. Our job is to protect them at all costs, during convoys and while they are at their venues. As a sniper, I help with the personal security, and when we reach our destination, I take an elevated position. The major benefit of this job, when the money is taken away, is only self-satisfaction. The best thing about being a sniper during missions is being able to watch over the rest of your team, and knowing that you're able to protect them.

The act of sniping for the military and for the State Department is very different. This job is fun because you are away from the big military force and are truly an independent operator. Here, there is no one to look over your shoulder, and the decisions that you make rely entirely upon your history and experience. This is why to become a contractor requires many years of experience. Also, the more experience, the better quality of jobs you may receive.

Another very important, and developing, aspect to sniping is technology. With the advancement of sniper operations in all sectors, technology is also rapidly growing to suit the skill. Recently there has been a development of advanced weapons, bringing sniping to a new level. One company, Space Dynamics Laboratory, from Utah, has developed the ARSS (Autonomous Rotorcraft Sniper System). Essentially, it is remote-controlled flying sniper system firing .338-caliber ammunition. The use of this is undeniably exciting for troops on the ground. A robot with the ability to collect intelligence, observe, report, and provide accurate fire, all with no cost to human lives, shows the rapidly changing world of sniping.

Other developments have come in many forms. Today, U.S. marines, soldiers, sailors, and airmen are being supplied with better weapons, optics, and equipment than ever before. Even more important, however, U.S. military snipers are now receiving the best training available, and the U.S. Army Sniper Association has helped immensely with their International Sniper Competition.

Every year, at Fort Benning, Georgia, military snipers gather for this competition. Sniper teams from all services, including law enforcement agencies, as well as sniper teams from various other countries, are invited to compete in the most comprehensive and realistic sniper training available. The event also brings snipers together to share valuable tactical information, relevant to any sniper, in any scenario.

In the years of the War on Terror, military snipers have made all the difference. Gains in technology and equipment have been phenomenal. At the end of the day, however, there will never be a substitute for an intelligent warrior equipped with a precision-fired weapon and the training to employ it as a sniper.